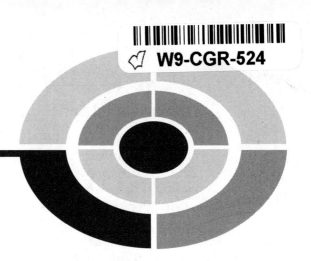

W9-CGR-524

Electricity Demystified

Demystified Series

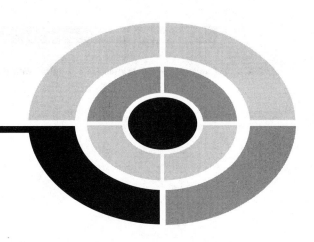

Electricity Demystified

STAN GIBILISCO

McGRAW-HILL
New York   Chicago   San Francisco   Lisbon   London
Madrid   Mexico City   Milan   New Delhi   San Juan
Seoul   Singapore   Sydney   Toronto

Cataloging-in-Publication Data is on file with the Library of Congress.

1 2 3 4 5 6 7 8 9 0   DOC/DOC   0 1 0 9 8 7 6 5

ISBN   0-07-143925-0

The sponsoring editor for this book was Judy Bass and the production supervisor was Pamela A. Pelton. It was set in Times Roman by Keyword Publishing Services. The art director for the cover was Margaret Webster-Shapiro. Cover design by Handel Low.

Printed and bound by RR Donnelley.

 This book was printed on recycled, acid-free paper containing a minimum of 50% recycled, de-inked fiber.

McGraw-Hill books are available at special quantity discounts to use as premiums and sales promotions, or for use in corporate training programs. For more information, please write to the Director of Special Sales, McGraw-Hill Professional, McGraw-Hill, Two Penn Plaza, New York, NY 10121-2298. Or contact your local bookstore.

To Samuel, Tim, and Tony from Uncle Stan

Mon. 30-April 2018
L C Libr. Rte. 30 Main

Electricity
Demystified $ 0.00 ~~~~

©2005

19.95

CONTENTS

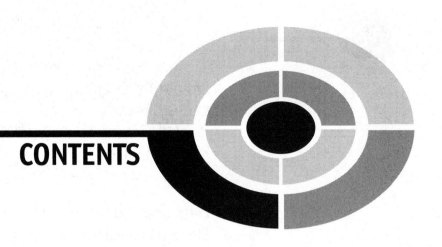

CONTENTS

Final Exam   289

APPENDIX 1   Answers to Quiz, Test,
              and Exam Questions   309

APPENDIX 2   Symbols Used in
              Schematic Diagrams   313

              Suggested Additional References   330

              Index   331

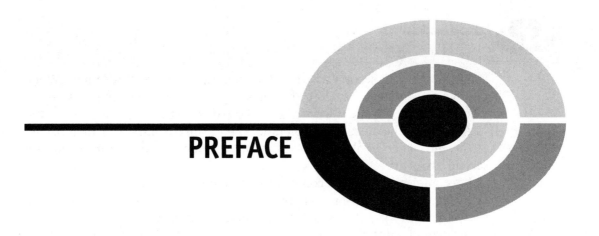

PREFACE

This book is for people who want to get acquainted with the concepts of elementary electricity and magnetism, without taking a formal course. It can serve as a supplemental text in a classroom, tutored, or home-schooling environment. It can also be useful for career changers who want to become familiar with basic electricity and magnetism.

This course is for beginners, and is limited to elementary concepts. The treatment is mostly qualitative. There's some math, but it doesn't go deep. If you want to study electronics following completion of this book, *Electronics Demystified* and *Teach Yourself Electricity and Electronics*, also published by McGraw-Hill, are recommended.

This book contains many practice quiz, test, and exam questions. They are all multiple-choice, and are similar to the sort of questions used in standardized tests. There is an "open-book" quiz at the end of every chapter. You may (and should) refer to the chapter texts when taking them. When you think you're ready, take the quiz, write down your answers, and then give your list of answers to a friend. Have the friend tell you your score, but not which questions you got wrong. Stick with a chapter until you get most of the answers correct.

This book is divided into sections called "Parts." At the end of each section is a multiple-choice test. Take these tests when you're done with the respective sections and have taken all the chapter quizzes. The section tests are "closed-book," but the questions are not as difficult as those in the quizzes. A satisfactory score is 75%. There is a final exam at the end of this course. It contains questions from all the chapters. Take this exam when you have finished all the sections and tests. A satisfactory score is at least 75%.

With the section tests and the final exam, as with the quizzes, have a friend tell you your score without letting you know which questions you missed. That way, you will not subconsciously memorize the answers. You might want to take each section test, and the final exam, two or three times. When you have got yourself a score that makes you happy, you can check to see where your knowledge is strong and where it is weak.

Answers to the quizzes, section tests, and the final exam are in an appendix at the end of the book. A table of circuit-diagram symbols is included in a second appendix.

I recommend that you complete one chapter a week. An hour or two everyday ought to be enough time for this. When you're done with the course, you can use this book as a permanent reference.

Incidentally, don't fly a kite in, or near, a thundershower to demonstrate that lightning is a form of electricity. Ben Franklin did that experiment a long time ago (according to popular legend, at least) and escaped with his life. Some Russian scientists tried it (for real) and they got killed.

Illustrations in this book were generated with *CorelDRAW*. Some of the clip art is courtesy of Corel Corporation.

Suggestions for future editions are welcome.

Stan Gibilisco

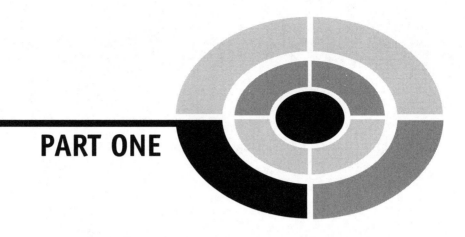

PART ONE

DC Electricity

CHAPTER 1

A Circuit Diagram Sampler

When learning how to read circuit diagrams, nothing beats practice. Reading circuit diagrams is like learning how to drive. You can read books about driving, but when it is time to get onto the road, you need practice before you feel comfortable. This chapter will get you into a frame of mind for diagram-reading. As you proceed along this course, you'll get better at it.

Block Diagrams

The *block diagram* is the easiest type of a circuit diagram to understand. The major components or systems are shown as rectangles, and the interconnecting wires and cables are shown as straight lines. The specialized components have unique symbols that are the same as those used in the more detailed circuit diagrams.

WIRES, CABLES, AND COMPONENTS

Figure 1-1 is a block diagram that shows an electric generator connected to a motor, a computer, a hi-fi stereo system, and a television (TV) receiver. Each major component is illustrated as a rectangle or a "block."

The interconnecting wires in this system are actually 3-wire electrical cords or cables. They appear as single, straight lines that run either vertically or horizontally on the page. In the interest of neatness, the symbols for wires and cables should be drawn only from side-to-side or up-and-down, following the "north-south/east-west" paths like the streets in a well-planned city on a flat land. A diagonal line should be used only when the diagram gets so crowded that it isn't practical to show a particular section of a wire or cable as a vertical or a horizontal line.

ADDING MORE ITEMS

The scenario in Fig. 1-1 is simple. None of the lines cross each other. But, suppose we decide to connect an alternating-current (AC) voltmeter to the circuit, and make it switchable so that it can be connected between an earth ground and the input of any one of the four devices receiving power? This will make the diagram more complicated, and we will want to let some of the lines cross.

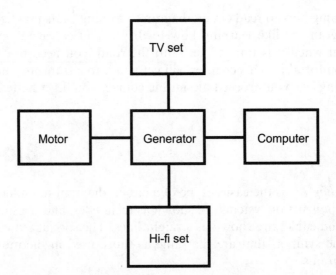

Fig. 1-1.   A simple block diagram of an electric generator connected to four common appliances.

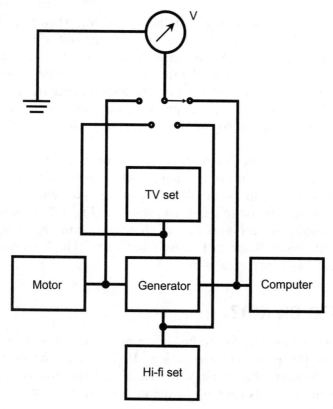

Fig. 1-2. Addition of a voltmeter, switch, and ground to an electrical system.

Figure 1-2 shows how we can illustrate the addition of a voltmeter (symbolized by a circle with an arrow and labeled V), together with a four-way switch (the symbol for which should be obvious). The voltmeter has two distinct terminals, one connected to the ground (the symbol with the three horizontal lines of different lengths) and the other connected to the central pole of the switch.

A LIMITATION

A block diagram does not, and is not intended to, portray all the details about a circuit. We don't know whether the hi-fi set in Fig. 1-1 or Fig. 1-2 is simple or sophisticated. We don't know what sort of features the computer has. Nor are we told whether the TV set is connected to cable, a satellite system, or a coat-hanger antenna! These details are not shown in the system block diagrams like Figs. 1-1 or 1-2.

It is not always clear how many conductors a cable, when represented by a single line, actually has. There are two lines coming out of the voltmeter in Fig. 1-2; each line represents a single-conductor wire. But among the generator and the four major appliances, the interconnecting lines represent 3-wire cords, not single-conductor wires. The lines running from the switch to the inputs of each of the major appliances, however, do represent single-conductor wires, connected to the voltage-carrying wire or "hot" in each of the four 3-conductor cords.

Why can't we show the 3-wire cords as sets of three lines, all parallel to each other? We can! But one of the main assets of block diagrams is the fact that they show things in as simple a manner as possible. In a complete, detailed schematic diagram of the system shown in Fig. 1-2, we would need to show the 3-conductor cords as sets of three lines running alongside each other. But in a block diagram this isn't necessary. The lines show general electrical paths, not individual wires.

CONNECTED OR NOT?

When two wires or cables are to be shown as connected, it is customary to put a dot at the point of intersection. In Fig. 1-2, the dots represent connections between the single wires running from the switch and the "hot" wires in the cords running to the motor, the computer, the hi-fi set, and the TV set.

What about the points where the lines representing wires from the switch cross lines representing the cords between the appliances and the generator, but there are no dots? The absence of a dot means that the wires or cables are not connected. It's as if they're both lying on the floor, insulated from each other, even though they happen to cross one over the other.

Figure 1-3A shows two lines crossing, but representing wires or cables that are not connected. Figure 1-3B shows lines representing two wires or cables that cross, but are connected. Figure 1-3C is a better way to show two wires or cables that cross and are connected to each other. This arrangement is better than the one at B, because the little dots in the diagrams are easy to overlook, and sometimes it looks like there's a dot at someplace, when there really isn't any. We shouldn't have to sit there and squint and ask ourselves, "Is there a dot at this point, or not?"

When three lines come together at a point, it usually means that they are connected, even if there is no dot at the point of intersection. Nevertheless, it's always a good diagram-drawing practice to put a heavy black dot at any point where wires or cables are meant to be connected.

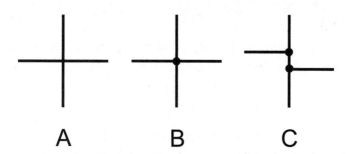

Fig. 1-3.   At A, the two wires are not connected. At B and C, they are. The rendition at C is preferable.

? PROBLEM 1-1

To which appliance power input is the voltmeter in Fig. 1-2 connected?

✔ SOLUTION 1-1

Note the position of the arrowed line inside the switch symbol. It runs to the terminal whose line goes to the line representing the power cord for the computer. Thus, the voltmeter in Fig. 1-2 is connected to the computer power input.

? PROBLEM 1-2

What is another way to show that two crossing lines represent wires or cables that are not connected?

✔ SOLUTION 1-2

A little "jump" or a "jog" can be added to one of the lines near the crossing point, as shown in Fig. 1-4.

Fig. 1-4.   Illustration for Problem 1-2.

? **PROBLEM 1-3**

Suppose we want to show the direction in which the electrical energy or power "moves" in the system shown by Fig. 1-2. How can we do that?

✔ **SOLUTION 1-3**

We can add arrows pointing outward from the generator toward each of the major appliances, as shown in Fig. 1-5. This indicates that "something flows" from the generator to these other devices. It's intuitively apparent that energy or power is the "something" that "flows" in this example. No arrows are added for the voltmeter, because an ideal voltmeter doesn't consume any power.

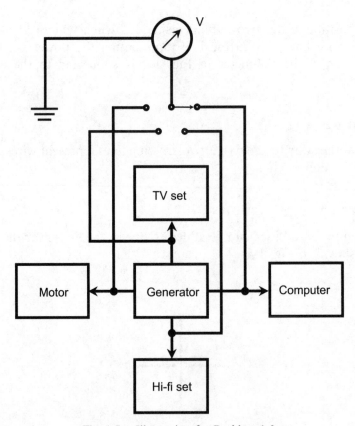

Fig. 1-5.   Illustration for Problem 1-3.

Schematic Diagrams

In this section, we'll look at a few basic schematic symbols that are often used in electricity (as opposed to electronics, where there are a lot more symbols). Let's examine a few commonplace electrical devices. Appendix 2 is a comprehensive table of symbols used to represent components in the electrical and the electronic systems. It is a good idea to start studying it now, and to review it often. When you're done with this course, you can use this appendix as a permanent reference.

FLASHLIGHT

A *flashlight* consists of a battery, a switch, and a light bulb. The switch is connected so that it can interrupt the flow of current through the bulb. Figure 1-6A shows a flashlight without a switch. The switch is added in Fig. 1-6B.

Note that the switch is connected in *series* with the bulb and the battery, rather than across (in *parallel* with) the bulb or the battery. The current must flow through all the three devices—the switch, the bulb, and the battery—in order to light up the bulb.

? **PROBLEM 1-4**

Is the flashlight bulb in the circuit shown by Fig. 1-6B illuminated? If so, why? If not, why not?

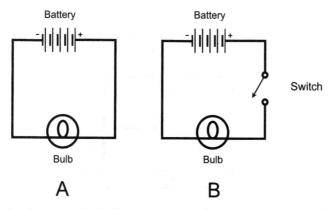

Fig. 1-6.   At A, a battery and a bulb connected together. At B, a switch is added, forming a common flashlight. This drawing is also the subject of Problem 1-4.

✔ **SOLUTION 1-4**

As shown in this diagram, the bulb is not illuminated because the switch is shown in the open position (not conducting).

? **PROBLEM 1-5**

What will happen if the switch is placed in parallel with the light bulb, as shown in Fig. 1-7?

✔ **SOLUTION 1-5**

The bulb will light up if the switch is open, as shown in Fig. 1-7. If the switch is closed, however, the bulb will go out because it is short-circuited. This action will also short out the battery, and that's a bad thing to do! A direct short circuit across a battery can cause chemicals to boil out of the battery. Some batteries can even explode when shorted out, and if the battery is large enough, the circuit wires can get so hot that they melt or start a fire.

VARIABLE-BRIGHTNESS LANTERN

Suppose we find an electric lantern bulb that is designed to work at 6 volts DC (6 V). It will light up with less voltage than that, but its brightness, in such a case, will be reduced. Figure 1-8 shows a 6 V battery connected to a 6 V bulb through a variable resistor called a *potentiometer*. The zig-zags in the symbol mean that the component is a resistor, and the arrow means that the resistance can be adjusted.

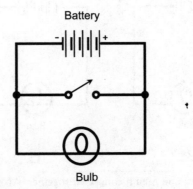

Fig. 1-7. Illustration for Problem 1-5.

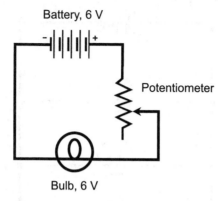

Fig. 1-8.   A variable-brightness lantern.

When the potentiometer is set for its lowest resistance, which is actually a direct connection, the bulb lights up to full brilliance. When the potentiometer is set for its highest resistance, the bulb is dim or dark. When the potentiometer is at intermediate settings, the bulb shines at intermediate brilliance.

[?] **PROBLEM 1-6**

What will happen if the potentiometer is connected in parallel with the light bulb and battery, rather than in series with it?

[✔] **SOLUTION 1-6**

With the potentiometer at its maximum resistance, the bulb will shine at its brightest. As the resistance of the potentiometer is reduced, the bulb will get dimmer, because the potentiometer will rob some of the current intended for the bulb. Depending on the actual value of the potentiometer and the amount of power it is designed to handle, the potentiometer will heat up or burn up if the resistance is set low enough. There will also be a risk of having the same problems as is the case when the battery is shorted out.

MULTIPLE-BULB CIRCUIT

Now, suppose we want to connect several bulbs, say five of them, across a single battery and have each of them receive the full battery voltage. We can do this by arranging the bulbs and the battery, as shown in the diagram of Fig. 1-9. This sort of a circuit, used with a 12 V battery, is commonly used

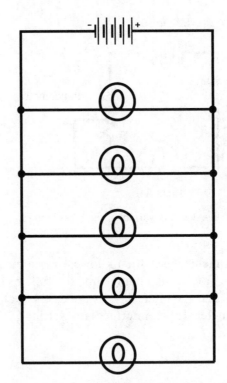

Fig. 1-9. A 5-bulb circuit. Symbols aren't labeled, because we already know what they represent.

in cars, boats, and campers. There are no switches shown in this circuit, so all the bulbs are illuminated all the time.

In this diagram, the symbols are not labeled. That's because you know, by now, what they represent anyway. In the standard operating practice, a schematic symbol is rarely labeled, unless an engineer or a technician might be confused without the label.

[?] **PROBLEM 1-7**

If one of the light bulbs in the circuit of Fig. 1-9 burns out, what will happen?

[✔] **SOLUTION 1-7**

The other bulbs will all remain shining, because they will still get the full battery voltage.

? **PROBLEM 1-8**

If one of the light bulbs in the circuit of Fig. 1-9 shorts out, what will happen?

✔ **SOLUTION 1-8**

In this case, the battery will also be shorted out. All the bulbs will go dark, because the short circuit will consume all the available battery current.

? **PROBLEM 1-9**

How can a switch be added to the circuit of Fig. 1-9, so that all the light bulbs can be switched on or off at the same time?

✔ **SOLUTION 1-9**

Figure 1-10 shows how this is done. The switch is placed right next to the battery, so that when it is opened, it interrupts the electrical path to all the bulbs. Here, the switch is placed next to the positive battery terminal. It could just as well be on the other side of the battery, next to the negative terminal.

? **PROBLEM 1-10**

How can switches be added to the circuit of Fig. 1-9, so that each light bulb can be individually switched on or off at the same time?

✔ **SOLUTION 1-10**

Figure 1-11 shows how this is done. There are five switches, one right next to each bulb. When a particular switch is opened, it interrupts the electrical path to the bulb to which it's directly connected, but does not interrupt the path to any of the other bulbs.

More Diagrams

There are plenty of things we can do with a battery, some light bulbs, some switches, and some potentiometers. The following paragraphs should help you get used to reading schematic diagrams of moderate complexity.

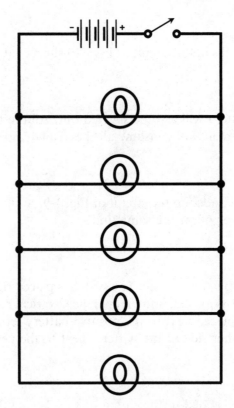

Fig. 1-10.   Illustration for Problem 1-9.

UNIVERSAL DIMMER

We can add a potentiometer to the circuit in Fig. 1-11, so that the brightnesses of all the bulbs can be adjusted simultaneously. In Fig. 1-12, the potentiometer acts as a universal light dimmer. The electricity, which follows the wires (straight lines), must pass through the potentiometer exactly once to go from the battery through any single bulb, and back to the battery again.

INDIVIDUAL DIMMERS

Figure 1-13 shows a circuit similar to the one in Fig. 1-11, except that in this case, each bulb has its own individual potentiometer. Therefore, the brightness of each light bulb can be adjusted individually.

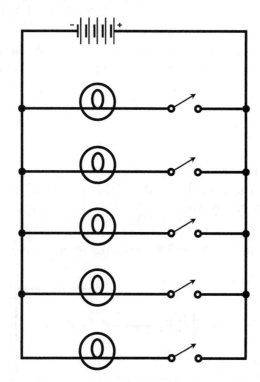

Fig. 1-11.   Illustration for Problem 1-10.

[?] **PROBLEM 1-11**

If one of the potentiometers, say the second from the top, is adjusted in the circuit of Fig. 1-13, it will affect the brilliance of the corresponding bulb. What about the brilliance of the other bulbs? Will the adjustment of the second potentiometer from the top affect the brightness of, say, the second bulb from the bottom?

[✔] **SOLUTION 1-11**

No. Each potentiometer in Fig. 1-13 affects the brilliance of its associated bulb, but not the others. The dimmers in this circuit are all independent.

[?] **PROBLEM 1-12**

Can we do anything to the circuit of Fig. 1-13, so that all the lights can be dimmed simultaneously, as well as independently?

☑ **SOLUTION 1-12**

Yes. We can add a potentiometer as in the scenario of Fig. 1-12, in addition to the five that already exist in Fig. 1-13. The result is shown in Fig. 1-14.

Quiz

This is an "open book" quiz. You may refer to the text in this chapter. A good score is 8 correct answers. All of the questions in this quiz refer to Fig. 1-15. Answers are in Appendix 1.

   1.   The circuit shown by Fig. 1-15 contains a battery, three motors, an ammeter (labeled A, which measures electric current), a lamp, five

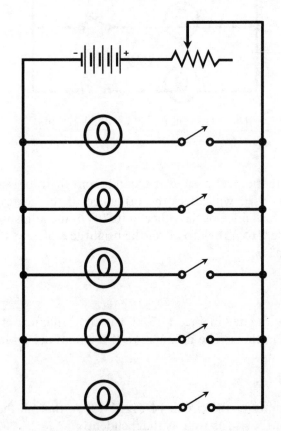

Fig. 1-12.   A circuit in which five light bulbs are individually switched, and are dimmed simultaneously.

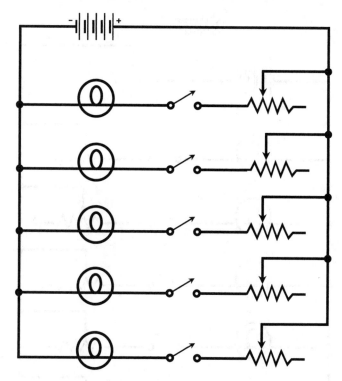

Fig. 1-13. A circuit in which five light bulbs are individually switched, and can be dimmed individually and independently. This drawing is also the subject of Problem 1-11.

potentiometers, and six switches. As shown in the diagram, some of the devices receive current, while others don't. Which devices are receiving current?

(a) All the three motors.
(b) Motor 1, the ammeter, and the lamp.
(c) Motors 2 and 3.
(d) None of the devices are receiving current.

2. What will happen if switch T is opened, but all the other switches are left in the same positions, as shown in Fig. 1-15?

(a) None of the devices will receive current.
(b) All of the devices will receive current.
(c) The devices that are now receiving current will not receive it, and the devices that are not receiving current will now receive it.
(d) It is impossible to predict.

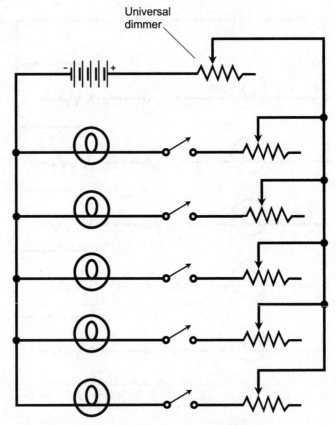

Fig. 1-14. Illustration for Problem 1-12. The potentiometer that serves as the universal dimmer is so labeled. All the unlabeled potentiometers are independent dimmers.

3. Suppose potentiometer X is adjusted. Which component(s), if any, will this affect, if the switches are in the positions shown?

(a) Motor 2.
(b) All of the components.
(c) None of the components.
(d) It is impossible to predict.

4. Suppose switch W is closed, and then potentiometer X is adjusted. Which component(s), if any, will this affect?

(a) Motor 2.
(b) All of the components.
(c) None of the components.
(d) It is impossible to predict.

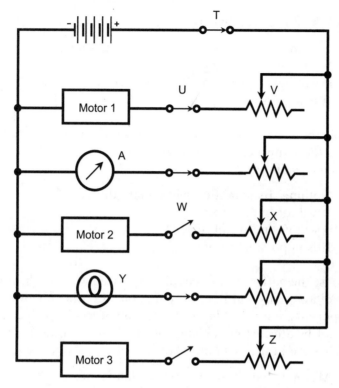

Fig. 1-15.   Illustration for Quiz Questions 1 through 10.

5. What will happen if switch U is opened?

(a) All the motors will fail to run, the meter will show zero current, and the lamp will go out.
(b) All of the components will run.
(c) Motor 1 will start running, but none of the other components will be affected.
(d) Motor 1 will stop running, but none of the other components will be affected.

6. Which component, if any, does potentiometer Z affect, assuming that the switches are in the positions shown by Fig. 1-15?

(a) All of them.
(b) None of them.
(c) Motor 3 only.
(d) All the components, except Motor 3.

7. If switch W is closed, what will happen to lamp Y?

   (a) Nothing. Its condition will remain the same.
   (b) It will go out, having been lit before.
   (c) It will light up, having been out before.
   (d) It is impossible to predict.

8. If potentiometer V is set so that its resistance is extremely high, making it the equivalent of an open switch, what will happen to the ammeter reading?

   (a) Nothing. Its condition will remain the same.
   (b) It will drop to zero.
   (c) It will go up to full scale.
   (d) It is impossible to predict.

9. Suppose the battery in this circuit is replaced with one having a different voltage. This will have an effect on the behavior of some of the components, but it will have no effect on others. Which component(s) will not be affected by a change in the battery voltage, assuming the switches are in the positions shown?

   (a) Motor 1, the ammeter, and the lamp.
   (b) Motor 1 only.
   (c) The lamp only.
   (d) Motors 2 and 3.

10. If switch T is opened and all the other switches are closed, which component(s), if any, will receive current?

   (a) All of them.
   (b) Only the motors.
   (c) None of them.
   (d) It is impossible to predict.

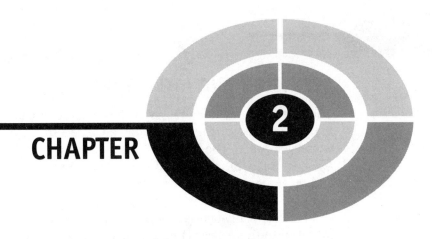

CHAPTER 2

Charge, Current, Voltage, and Resistance

What is electricity? Why can it do so many things when a circuit is closed, and yet seem useless when a circuit is open? In this chapter we'll investigate the nature of electricity.

Charge

In order for electricity to exist, there must be a source of *electric charge*. There are two types of charges. Scientists chose the terms *positive* and *negative* (sometimes called *plus* and *minus*) to represent the two kinds of charges.

REPULSION AND ATTRACTION

The first experimenters observed that when two electrically charged objects are brought close, they are either attracted to each other or else repelled from each other. The force of attraction or repulsion, called the *electrostatic force*, operates through empty space between objects, just as magnets attract and repel, depending on the way they are positioned. But electrostatic force obviously isn't the same phenomenon as magnetic force.

Two electrically charged objects attract if one has a positive charge and the other a negative charge (Fig. 2-1A). If both the objects are positively charged (Fig. 2-1B) or both the objects are negatively charged (Fig. 2-1C), they repel. The magnitude of the force, whether it is attraction or repulsion, depends on two factors:

- The total extent to which the objects are charged
- The distance between the centers of the objects

As the total extent of the charge, also called the *charge quantity* (considered on the two objects taken together) increases, and if the distance between the centers of the objects does not change, the force between the objects increases in direct proportion to the total charge (Fig. 2-2). As the separation between the centers of the two charged objects increases, and if the total charge remains constant, the force decreases according to the square of the distance (Fig. 2-3).

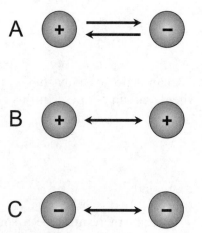

Fig. 2-1. At A, two objects having opposite charges attract. At B and C, two objects having similar charges repel.

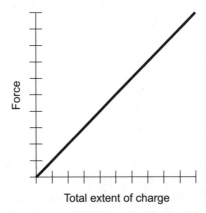

Fig. 2-2.   When the total quantity of charge on two objects increases but nothing else changes, the force between them increases in direct proportion to the total charge.

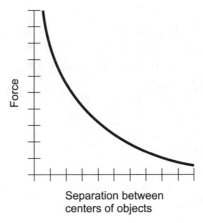

Fig. 2-3.   When the distance between the centers of two charged objects increases but nothing else changes, the force between them decreases according to the square of the distance.

THE ATOM

Scientists have known for a long time that all matter is not continuous, but exists in tiny pieces or particles. The more we study matter, the more complicated things seem to get. There are particles that behave like matter in some situations, and like energy in others. Sometimes things act like particles and sometimes they act like waves. In order to understand the basics of electricity, let's take a highly simplified view.

Matter is made up of particles called *atoms*. These atoms are in turn made up of smaller particles called *protons*, *neutrons*, and *electrons*. Protons and neutrons are incredibly tiny and dense. They tend to be lumped together at the centers of atoms. The center of an atom is called the *nucleus* (plural: *nuclei*). Electrons are much less dense than protons or neutrons, and they move around a lot more. Some electrons "orbit" a specific nucleus and stay there indefinitely. But it's not uncommon for an electron to move from one atomic nucleus to another. Protons and neutrons, which actually compose the nucleus, rarely move from one atom to another.

Protons and electrons carry equal and opposite electric charges. By convention, protons are considered electrically positive, and electrons, electrically negative. The amount of charge on any proton is the same as the amount of charge on any other proton. Similarly, the amount of charge on any electron is the same as the amount of charge on any other electron. Neutrons don't carry any electrical charge.

ELECTRONS

Electrons, which are sometimes "loyal" to specific nuclei but often "free-roaming," are of particular interest in the study of electricity. An excess or a deficiency of electrons on an object gives that object a *static electric charge*, also called an *electrostatic charge*. If an object contains more number of total electrons than the total protons, then that object is said to be *negatively charged*. If an object contains fewer number of total electrons than the total protons, then that object is *positively charged*.

In charged objects of reasonable size, the number of electrons involved is huge. When you shuffle across a carpeted room on a dry day, your body acquires an electrostatic charge that consists of millions upon millions of electrons that either get accumulated or drawn out. When you imagine this, you might wonder how all those electrons can build up or run short on your body without putting your life in danger.

Under certain conditions, electrostatic charge is harmless. But at times, such as when you stand in an open field during a thunderstorm, a truly gigantic charge—vastly greater than the charge you get from shuffling across a carpet—can build up on your body. If your body acquires a massive enough charge, and then the charge difference between your body and something else (such as a cloud) is suddenly equalized, you can get injured or even killed.

UNITS OF CHARGE

Two units are commonly employed to measure, or quantify, an electrical charge. The most straightforward approach is to consider the charge on a single electron as the equivalent of one electrical charge unit. This charge is the same for every electron we observe under ordinary circumstances. The quantity of charge on a single electron is called an *elementary charge* or an *elementary charge unit.*

An object that carries one elementary charge unit (ECU), either positive or negative, is practically impossible to distinguish from an electrically neutral object. The ECU is an extremely small unit of charge. More often, a unit called the *coulomb* is used. One coulomb is approximately 6,240,000,000,000,000,000 ECU. This big number is written using *scientific notation*, also called the *power-of-10 notation*, as 6.24×10^{18}. The word "coulomb" or "coulombs" is symbolized by the non-italic, uppercase letter C. Thus:

$$1\,C = 6.24 \times 10^{18}\,ECU$$

$$2\,C = 2 \times 6.24 \times 10^{18} = 1.248 \times 10^{19}\,ECU$$

$$10\,C = 10 \times 6.24 \times 10^{18} = 6.24 \times 10^{19}\,ECU$$

$$0.01\,C = 0.01 \times 6.24 \times 10^{18}\,ECU = 6.24 \times 10^{16}\,ECU$$

In this book, we won't use scientific notation very often. If you're serious about studying electricity or any such scientific discipline, you ought to get comfortable with scientific notations. For the time being, it's good enough for you to remember that 1 C represents a moderate amount of electrical charge, something often encountered in the real world.

? **PROBLEM 2-1**

Imagine two charged spherical objects. Assume that the charge is distributed uniformly throughout either sphere. Suppose the left-hand sphere contains 1 C of positive charge, and the right-hand sphere contains 1 C of negative charge (Fig. 2-4A). This results in an attractive electrostatic force, *F*, between the two spheres. Now suppose the charge on either spheres is doubled, but the distance between their centers does not change. What happens to the force?

✔ **SOLUTION 2-1**

The force is quadrupled to 4*F*. The total charge quantity, represented by the product of the charges on the two objects, increases by a factor of $2 \times 2 = 4$, as shown in Fig. 2-4B.

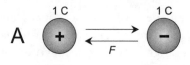

Fig. 2-4. Illustration for Problem 2-1. At A, the total charge is 1 C, resulting in a force of F. At B, the total charge is 4 C while the separation distance is unchanged, resulting in a force of $4F$.

? PROBLEM 2-2

Imagine the same two charged spherical objects as in the previous problem. Assume, again, that the charge is distributed uniformly throughout either sphere. The left-hand sphere contains 1 C of positive charge, and the right-hand sphere contains 1 C of negative charge (Fig. 2-5A). This results in an attractive electrostatic force, F, between the two spheres. Suppose the charge on either sphere is doubled, and the distance between their centers is also doubled. What happens to the force?

✔ SOLUTION 2-2

The force is unchanged. Doubling the charge on either sphere, but not changing the distance, increases the force by a factor of 4, as we have seen. But increasing the distance between the centers of the spheres causes the force to diminish by a factor equal to the square of the increase, which is a factor of $2^2 = 4$. In this case, the increase in force caused by the charge

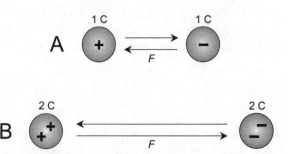

Fig. 2-5. Illustration for Problem 2-2. At A, the total charge is 1 C, resulting in a force of F. At B, the total charge is 4 C, and the separation distance is doubled. The force is still equal to F.

increase is canceled out by the decrease in force caused by the increase in the separation.

Current

When charged particles move, the result is *electric current*. Usually, the particles are electrons. But any charged object, when it moves, produces current. In a vacuum, protons can travel at a high speed under certain conditions. Atomic nuclei, consisting of protons and neutrons "stuck together in a clump," can also travel through a vacuum. These moving charge carriers produce current, although not of the sort you encounter every day!

CONDUCTORS

In certain materials, electrons can move easily from one atomic nucleus to another. In other materials, the electrons move, but with difficulty. In some materials, it is almost impossible to get electrons to move among the nuclei. An electrical *conductor* is a substance in which the electrons are highly mobile.

The best conductor among the common materials at room temperature is pure elemental silver. Copper and aluminum are also excellent electrical conductors. Iron, steel and most other metals are fair to good conductors of electricity. Some liquids are good conductors. Mercury is one example. Salt water is a fair conductor, but pure distilled water is a poor conductor. Gases are, in general, poor conductors, because the atoms or molecules are too far apart to allow a free exchange of electrons.

Electrons in a conductor do not move in a steady stream like the molecules of water through a garden hose, although this is not a bad analogy in some contexts. In an electrical conductor, the electrons "jump" between the adjacent atomic nuclei (Fig. 2-6). This happens to countless atoms throughout a substance when it conducts electricity. As a result, millions of electrons pass a given point each second in an electrical circuit that carries current.

INSULATORS

An electrical *insulator* is a substance in which electrons do not readily pass from atom to atom. Under ordinary conditions, insulators prevent current from flowing. Insulators are therefore, poor conductors.

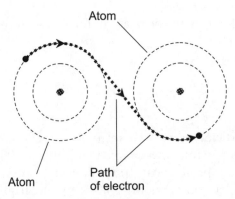

Atom

Atom

Path
of electron

Fig. 2-6. In an electrical conductor, electrons pass easily from atom to atom. This drawing is
greatly simplified.

Most gases are good electrical insulators. Glass, dry wood, paper, and
plastics are other examples. Pure water is a good electrical insulator,
although it conducts some current when minerals are dissolved in it. Certain
metallic oxides can be good insulators, even if the metal in pure form is a
good conductor. Aluminum oxide is an example.

An insulating material is sometimes called a *dielectric*. This term arises
from the fact that it keeps electrical charges apart, preventing the flow
of electrons that would otherwise occur between two objects or locations in
a circuit. When there are two separate regions of electric charge having
opposite polarity that are kept apart by a dielectric, that pair of charges is
called an *electric dipole*.

THE AMPERE

Current is measured in terms of the number of *charge carriers* that pass a
single point in one second (1 s). Usually, a great many charge carriers
go past any given point in 1 s, even if the current is small. For this reason,
current is rarely expressed directly in ECU per second (ECU/s), but instead
is expressed in coulombs per second (C/s). A current of 1 C/s is called an
ampere (symbolized A), and this is the standard unit of electric current.

In physics, electrical current is considered to flow from the positive to
the negative pole. This is known as *theoretical current* or *conventional current*.
If you connect a light bulb to a battery, therefore, the theoretical current
flows out of the positive terminal and into the negative terminal. But the
electrons, which carry the charge, flow in the opposite direction, from negative to positive. This is the way engineers usually think about current.

? **PROBLEM 2-3**

Suppose 6.24×10^{15} electrons flow past a certain point in 1 s of time. What is the current?

✔ **SOLUTION 2-3**

Note that 6.24×10^{15} is exactly 0.001 C. This means that 0.001 C/s flow past the point. This is equivalent to 0.001 A of current.

Voltage

Current can flow only if charge carriers are "pushed" or "motivated." The "motivation" can be provided by a buildup of electrostatic charges, as in the case of a lightning stroke. When the charge builds up, with positive polarity (shortage of electrons) in one place and negative polarity (excess of electrons) in another place, a powerful *electromotive force* (EMF) exists. This force, also known as *voltage* or *electrical potential*, is expressed in *volts* (symbolized V). You might also hear voltage spoken of as a sort of "electrical pressure."

COMMON VOLTAGES

Household electricity in the United States, available at standard electrical outlets, has an effective voltage of between 110 V and 130 V; usually it is about 117 V (Fig. 2-7A). This is not a steady voltage; it changes polarity at a steady rate of several dozen times a second. A typical automotive battery has an EMF of 12 V (Fig. 2-7B). This is a steady, constant voltage. The charge that you acquire when walking on a carpet with hard-soled shoes can be several thousand volts. Before a discharge of lightning, millions of volts build up between the clouds, or between a cloud and the ground.

STATIC ELECTRICITY

It is possible to have an EMF without a flow of current. This is the case just before a lightning bolt occurs, and before you touch a grounded or large metal object after walking on the carpet. It is the case between the terminals

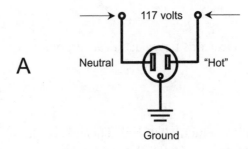

A

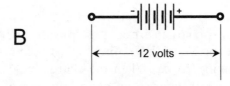

B

Fig. 2-7. At A, voltage at a common household utility outlet. At B, voltage at the terminals of an automotive battery.

of an electrical outlet when nothing is plugged into it. It is also true of a battery when there is nothing connected to it. When a voltage exists without current, it is called *static electricity*. "Static" in this context means "not moving." Current can flow only if a conductive path is placed between two points having different voltages.

Even a large EMF does not necessarily drive very much current through a conductive object. Consider your body after you have been shuffling around on a carpet. The voltage seems deadly if you think about it in terms of sheer EMF (thousands of volts), but not many coulombs of charge accumulate. Thus, not much current flows through your finger when you touch a grounded or massive metal object. You don't get a deadly shock.

CURRENT IS THE KILLER, BUT . . .

Perhaps you have heard that current, not voltage, kills people. This statement is true in a literal sense, but it is simplistic, and overlooks the way things work in the real world. It's like saying that if you stand on a the edge of a cliff you are safe; it is only after you jump off that you are in peril. High voltage is dangerous. Don't let anybody convince you otherwise. Even a moderate voltage can present an electrocution hazard. When you are around anything that carries significant voltage, you ought to have the same respect for it that you would have for a cliff as you walk along its edge.

When there are plenty of coulombs available, it doesn't take much voltage to produce a lethal flow of current, if a good path is provided for the flow of the charge carriers. This is why it is dangerous to repair an electrical device when it is plugged in or when power is otherwise applied. A "hot" utility outlet can pump charge carriers through your body fast enough, and for a long enough time, to kill you.

VOLTAGE VERSUS CURRENT

If a voltage is reasonably steady and doesn't alternate in polarity, the current through an electrical component varies in *direct proportion* to the voltage supplied to the component (Fig. 2-8), as long as the characteristics of the component do not change. This means that, if the voltage is doubled, the current doubles. If the voltage falls to 1/100 of its original value, so will the current.

This neat, straight-line relationship between voltage and current holds only as long as a component always conducts to the same extent. In many components, the electrical *conductance* changes somewhat as the current varies. This is the case, for example, in an electric bulb. The conductance is different when the filament carries a lot of current and glows white hot, as compared to when it is carrying only little current and not glowing at all. Some electrical components and devices are designed to have a constant

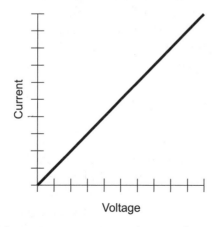

Fig. 2-8. When the voltage across a component increases but nothing else changes, the current through it increases in direct proportion to the voltage.

conductance, even when the voltages across them (and thus the currents through them) vary over a wide range.

☐? **PROBLEM 2-4**

Why is it dangerous to operate electrical appliances with bare feet on a wet pavement or on a bare ground, but much less dangerous if you are wearing rubber boots on your feet and leather gloves on your hands? The voltage is the same in both situations.

☑ **SOLUTION 2-4**

When you are bare-handed and aren't wearing proper shoes, it is possible for the household voltage to drive lethal current through your body from an appliance to the ground or a wet floor. If you wear hole-free, dry leather gloves along with hole-free, dry rubber boots, you are safer because the voltage cannot drive significant current through your body. (But you must still be careful. Before you work on any electrical appliance or system, always unplug the appliance or shut off the electricity completely at the fuse box or the breaker box.)

Resistance

Nothing conducts electricity with perfect efficiency. Even the best conductors have a little bit of *resistance*, which is opposition to the electric current. Silver, copper, aluminum, and most other metals have fairly low resistance. Some materials, such as carbon and silicon, have moderate resistance. Electrical insulators have high resistance. Resistance is the opposite of conductance, just as darkness is the opposite of light.

THE OHM

Resistance is measured in *ohms*. The word "ohm" or "ohms" is symbolized by the uppercase Greek letter, omega (Ω). The higher the value in ohms, the greater is the resistance, and the more difficult it is for current to flow. In an electrical system, it is usually desirable to have as low a resistance or an *ohmic value* as possible, because resistance converts electrical energy into heat. This heat is called *resistance loss* or *ohmic loss*, and in most cases represents energy wasted.

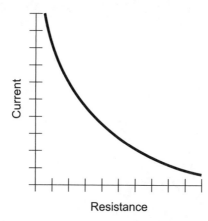

Fig. 2-9. When the voltage across a component remains constant, the current through it varies inversely with its resistance.

When 1 V of EMF exists across 1 Ω of resistance, there is a current of 1 A through the resistance (assuming the voltage source can deliver that much current). If the resistance doubles, the current is cut in half. If the resistance is cut in half, the current doubles. If the resistance increases by a factor of 5, then the current decreases to 1/5 its previous value. If the resistance is cut to 1/5 its previous value, then the current increases by a factor of 5. The current, given a constant voltage, varies in *inverse proportion* to the resistance (Fig. 2-9).

When current flows through a resistive material, it gives rise to a voltage, also called a *potential difference*, across the resistive object. The larger the current through the resistance, the greater is the voltage across it. In general, this voltage is directly proportional to the current through the resistive object (Fig. 2-10).

PRACTICAL RESISTANCE

Under normal circumstances, there is no such a thing as a substance with no resistance at all. No conductor is perfect in everyday life. Some materials can be considered perfect conductors in the absence of all heat—temperatures near *absolute zero*—but these are special exceptions. (The phenomenon in which conductors lose their resistance in extreme cold is known as *superconductivity*. We'll look at it in Chapter 12.)

Just as there is no such a thing as a perfectly resistance-free substance, there isn't a truly infinite resistance, either. Even distilled water or dry air

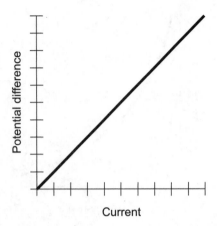

Fig. 2-10. When the current through a component increases but the resistance does not change, the potential difference across the component increases in direct proportion to the current.

conducts electricity to some extent, although the effect is usually so small that it can be ignored. In some applications, materials are selected on the basis of "how nearly infinite" the resistance is.

In a practical circuit, the resistances of certain components vary, depending on the conditions under which the components are operated. A transistor, for example, might have high resistance some of the time, and very low resistance at other times. This high/low fluctuation can be made to take place thousands, millions, or billions of times each second, an effect that makes possible the operation of high-speed switching devices of the sorts found in personal computers and home entertainment equipment.

? **PROBLEM 2-5**

Imagine an electrical device (let's call it component X) with a battery connected to it. Component X has a certain resistance. The voltage from the battery causes some current to flow through the component X. Now assume that the battery voltage is doubled, and the resistance of component X, cut in half. What happens to the current?

✔ **SOLUTION 2-5**

If the resistance of component X didn't change, then doubling the voltage would double the current. If the voltage across component X didn't change, then cutting the resistance in half would double the current. Because both things happen at once—the voltage is doubled and the resistance is cut in half—the current through component X is multiplied by a factor of 2, twice over. That means it is quadrupled.

Quiz

This is an "open book" quiz. You may refer to the text in this chapter. A good score is 8 correct answers. Answers are in Appendix 1.

1. If an object contains 1 C of electrical charge, and then that object is discharged at a constant rate so after 1 s it is electrically neutral, then

    (a) the voltage during the discharge process is 1 V.
    (b) the resistance during the discharge process is 1 Ω.
    (c) the current during the discharge process is 1 A.
    (d) the potential difference is 1 ECU.

2. If an object contains fewer number of total electrons than the number of total protons, then that object

    (a) is electrically charged.
    (b) has a high resistance.
    (c) contains electrical current.
    (d) cannot conduct electricity.

3. The voltage at a wall outlet is more dangerous than the voltage that builds up when you shuffle around on a carpet because

    (a) the utility outlet can send a higher current through your body.
    (b) the voltage at the wall outlet is much greater.
    (c) the resistance at the wall outlet is much higher.
    (d) the outlet contains more electrostatic force.

4. Examine Fig. 2-11. It shows a battery connected to a potentiometer and a lamp. Suppose the voltage of the battery (E) and the ohmic value of the potentiometer (R) are set such that the lamp glows partially, but not to its full brilliance. Suppose the setting of the

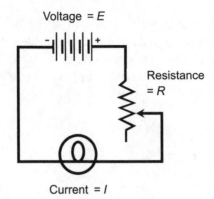

Voltage = E

Resistance = R

Current = I

Fig. 2-11.   Illustration for Quiz Questions 4 through 6.

potentiometer is adjusted such that R increases slightly, but E does not change. If this is done, and nothing else in the circuit changes, then

(a)   the bulb shines less brightly because the current (I) through it decreases.
(b)   the bulb shines more brightly because the current (I) through it increases.
(c)   the bulb shines less brightly because the voltage across it increases.
(d)   the bulb burns out because its resistance increases.

5.   If the value of E in the circuit of Fig. 2-11 is increased slightly but the value of R does not change, then the value of I

(a)   decreases slightly.
(b)   increases slightly.
(c)   decreases a lot.
(d)   increases a lot.

6.   Suppose that the lamp in Fig. 2-11 is shorted out, so that current does not flow through it, instead, only flows through the potentiometer. Under these conditions, if the value of R is doubled and the value of E is also doubled, then the current through the potentiometer

(a)   does not change.

(b)   doubles.

(c)   is cut in half.

(d)   increases, but more information is necessary to know how much.

7.   Imagine two charged objects, called X and Y. As they are brought closer to each other, assuming the charge remains constant on both the objects, the electrostatic force between them can be expected to

(a)   stay the same.

(b)   get stronger.

(c)   get weaker.

(d)   change from attraction to repulsion.

8.   "Electrical pressure" is a slang expression that refers to

(a)   any current that goes back and forth.

(b)   any current that does not change direction.

(c)   any current that passes through a high resistance.

(d)   electromotive force.

9.   Fill in the blank to make the following sentence true: "The _____ is a subatomic particle that contains no electrical charge of its own."

(a)   proton

(b)   electron

(c)   neutron

(d)   nucleus

10.   An electrical charge of 1 ECU is

(a)   equivalent to millions and millions of coulombs.

(b)   equivalent to 1 C.

(c)   equivalent to a tiny fraction of 1 C.

(d)   equivalent to 1 V.

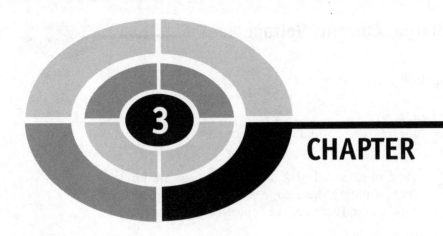

CHAPTER 3

Ohm's Law, Power, and Energy

In this chapter, we'll look more closely at the ways current, voltage, and resistance are related in DC circuits. We'll also learn how to calculate electrical power and energy, based on current, voltage, and resistance. But first, let's see how large and small fractions of units are expressed.

Prefix Multipliers

Scientists attach *prefix multipliers* to the words representing units, in order to express large multiples or small fractions of those units. Table 3-1 outlines the prefix multipliers commonly encountered in electricity.

Table 3-1   Prefix multipliers commonly used in electricity.

Designator	Symbol	Multiplier
pico-	p	0.000 000 000 001
nano-	n	0.000 000 001
micro-	μ or mm	0.000 001
milli-	m	0.001
kilo-	K or k	1000
mega-	M	1,000,000
giga-	G	1,000,000,000
tera-	T	1,000,000,000,000

? **PROBLEM 3-1**

A particular electrical component carries 3 μA of direct current. What is this current in amperes?

✔ **SOLUTION 3-1**

From the table, observe that the prefix "micro-" (μ) represents multiples of 0.000001 (millionths). Thus, 3 μA is equal to 0.000003 A.

? **PROBLEM 3-2**

A DC voltage source is specified as producing 2.2 kV. What is this in volts?

✔ **SOLUTION 3-2**

The table says that the prefix "kilo-" (k) stands for multiples of 1000 (thousands). Therefore 2.2 kV = 2.2 × 1000 V = 2200 V.

? **PROBLEM 3-3**

A resistor has a value of 47 MΩ. What is this in ohms?

✓ **SOLUTION 3-3**

From the table, you can see that the prefix "mega-" (M) stands for multiples of 1,000,000 (millions). Therefore, 47 MΩ = 47 × 1,000,000 Ω = 47,000,000 Ω.

Ohm's Law

In any active DC circuit, current flows through media having a certain amount of resistance. The current is "pushed" through the resistive medium by a voltage. The current, voltage, and resistance interact in a predictable and precise way. *Ohm's Law* defines this in mathematical terms.

THREE OHM'S LAW FORMULAS

Figure 3-1 shows a DC circuit containing a battery, a resistor, and an ammeter that measures the current through the resistor. Let E stand for the battery voltage (in volts), let I stand for the current through the resistor (in amperes), and let R stand for resistance of the resistor (in ohms). Three

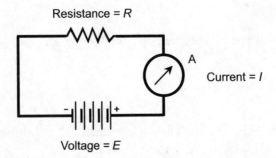

Fig. 3-1. A basic DC circuit for demonstrating current, voltage, and resistance calculations using Ohm's Law.

formulas denote Ohm's law:

$$E = IR$$

$$I = E/R$$

$$R = E/I$$

You should use units of volts (V), amperes (A), and ohms (Ω) for Ohm's Law to work. If quantities are given in units other than volts, amperes, and ohms, you must convert to these units and then calculate. After that, you can convert the units back to whatever you like.

SOME CALCULATIONS WITH OHM'S LAW

When you want to determine the current in a DC circuit, you must know the voltage and the resistance. Ohm's Law can also be used to determine an unknown voltage when the current and the resistance are known, or to determine an unknown resistance when the voltage and current are known.

For simplicity, let's be idealists and assume that the wires in the circuit of Fig. 3-1 conduct perfectly. That is, let's assume they have zero resistance. That way, we don't have to worry about their effect in the overall behavior of the circuit. (In real life, wires always have a little resistance, and this sometimes influences the behavior of components, devices, and systems.)

? **PROBLEM 3-4**

Suppose the battery in the circuit of Fig. 3-1 produces 6 V, and the resistor has a value of 2 Ω. What is the current?

✔ **SOLUTION 3-4**

Plug in the numbers to the Ohm's Law formula for current, as follows:

$$I = E/R$$

$$= (6/2)\text{A}$$

$$= 3\text{A}$$

That is a significant electrical current.

? **PROBLEM 3-5**

Suppose the battery in the circuit of Fig. 3-1 produces 6 V, but the resistance is 6 kΩ. What is the current?

✔ **SOLUTION 3-5**

Convert the resistance to ohms: 6 kΩ = 6 × 1000 Ω = 6000 Ω. Then plug the values in:

$$I = E/R$$

$$= (6/6000)A$$

$$= 0.001A$$

This is equivalent to a milliampere (1 mA), which is a small amount of current in most applications.

? **PROBLEM 3-6**

Suppose the resistor in the circuit of Fig. 3-1 has a value of 100 Ω, and the measured current is 10 mA. What is the DC voltage?

✔ **SOLUTION 3-6**

Use the formula $E = IR$. First, convert the current to amperes: 10 mA = 10 × 0.001 A = 0.01 A. Then plug in the numbers:

$$E = IR$$

$$= (0.01 \times 100)V$$

$$= 1V$$

That's a low voltage, a little less than that produced by a typical dry cell.

? **PROBLEM 3-7**

If the voltage is 24 V and the ammeter shows 3 A, what is the value of the resistor?

✔ **SOLUTION 3-7**

Use the formula $R = E/I$, and plug in the values directly, because they are expressed in volts and amperes:

$$R = E/I$$
$$= (24/3)\Omega$$
$$= 8\,\Omega$$

That would be regarded as a low resistance in some situations, but as a high resistance in others.

Power

Power is the rate at which energy is consumed or used. The standard unit of power is *watt*. The words "watt" and "watts" are abbreviated as an uppercase, non-italic letter W. When power is expressed as a variable in an equation, it is symbolized by an uppercase, italic P.

A conventional household bulb uses 25 W to 100 W. Sometimes larger units are used to express electrical power, such as the kilowatt (kW), equivalent to 1000 W, or the megawatt (MW), equivalent to 1,000,000 W. You will also occasionally hear about the milliwatt (mW), equivalent to 0.001 W, or the microwatt (μW), equivalent to 0.000001 W.

THREE POWER FORMULAS

Figure 3-2 shows a DC circuit containing a battery, a motor, a voltmeter that measures the voltage across the motor, and an ammeter that measures the current through the motor. Let E stand for the voltmeter reading (in volts), let I stand for the ammeter reading (in amperes), and let R stand for the resistance of the motor (in ohms). Three formulas can be used to determine the power, P (in watts) consumed by the motor in this circuit:

$$P = EI$$
$$P = E^2/R$$
$$P = I^2R$$

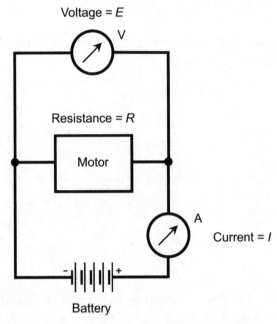

Fig. 3-2. A DC circuit for demonstrating power calculations. The power-consuming device is an electric motor.

As with Ohm's Law, you should use units of volts (V), amperes (A), and ohms (Ω) for power calculations to come out in watts. If quantities are given in units other than volts, amperes, and ohms, you must convert those quantities to these units before you calculate.

SOME POWER CALCULATIONS

In order to determine the power in a DC circuit, you must know two out of the three basic parameters (current, voltage, and resistance). As with the Ohm's law problems, let's assume here that the wires in the circuit conduct perfectly.

? **PROBLEM 3-8**

Suppose the voltmeter in the circuit of Fig. 3-2 reads 6 V, and the ammeter reads 5 A. What is the power consumed by the motor?

✔ **SOLUTION 3-8**

Plug in the numbers as follows:

$$P = EI$$
$$= (6 \times 5)\text{W}$$
$$= 30\text{W}$$

Note that the battery voltage is 6 V, because the voltmeter is connected directly across the battery, as well as directly across the motor.

? **PROBLEM 3-9**

Suppose the voltmeter in the circuit of Fig. 3-2 shows 12 V, and the resistance of the motor is 14.4 Ω. What is the power consumed by the motor?

✔ **SOLUTION 3-9**

Plug the values into the formula for power in terms of voltage and resistance:

$$P = E^2/R$$
$$= (12 \times 12/14.4) \text{ W}$$
$$= (144/14.4) \text{ W}$$
$$= 10 \text{ W}$$

? **PROBLEM 3-10**

Suppose the motor in the circuit of Fig. 3-2 has an internal resistance of 10 Ω, and the current measured through it is 600 mA. What is the power?

✔ **SOLUTION 3-10**

Use the formula $P = I^2R$. First, convert the current to amperes: 600 mA = 0.6 A. Then plug in the numbers:

$$E = I^2R$$
$$= (0.6 \times 0.6 \times 10) \text{ W}$$
$$= (0.36 \times 10) \text{ W}$$
$$= 3.6 \text{ W}$$

Energy

Energy consists of power expended or used over a period of time. The standard unit of energy is *joule*. The words "joule" and "joules" are abbreviated as an uppercase, non-italic letter J. A joule is a watt-second (W · s or Ws), which is the equivalent of one watt (1 W) expended for one second (1 s). When energy is expressed as a variable in an equation, let's symbolize it by an uppercase, italic letter W. (We can't use E because that already stands for the voltage variable!)

ENERGY UNITS

In electricity, energy is not often expressed in joules, because the joule is a tiny unit. More often it is expressed in watt-hours (W · h or Wh). A watt-hour is the equivalent of one watt (1 W) consumed for one hour (1 h) or 3600 seconds (3600 s). Therefore:

$$1\,\text{Wh} = 3600\,\text{Ws}$$
$$= 3600\,\text{J}$$

In medium-sized electrical systems, such as households, even the watt-hour is a small unit of energy. Then the kilowatt-hour (kW · h or kWh) is used. This is the equivalent of 1 kW (1000 W) expended for 1 h. Thus:

$$1\,\text{kWh} = 1000\,\text{Wh}$$
$$= (1000 \times 3600)\,\text{J}$$
$$= 3{,}600{,}000\,\text{J}$$

In large electrical networks such as the utility grids in cities, even the kilowatt-hour is a small unit! Then the megawatt-hour (kW · h or MWh) is used. This is the equivalent of 1 MW (1,000,000 W) expended for 1 h. Thus:

$$1\,\text{MWh} = 1{,}000{,}000\,\text{Wh}$$
$$= (1{,}000{,}000 \times 3600)\,\text{J}$$
$$= 3{,}600{,}000{,}000\,\text{J}$$

THREE ENERGY FORMULAS

Figure 3-3 shows a DC circuit containing a battery, an incandescent lamp, an ammeter that measures the current through the lamp, a switch to turn

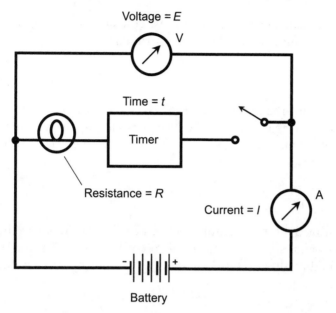

Voltage = E

Time = t

Resistance = R

Current = I

Battery

Fig. 3-3. A DC circuit for demonstrating energy calculations. The energy-consuming component is an incandescent bulb. This diagram is also the subject of Quiz Questions 1, 4, 5, and 6.

the lamp on and off, a timer that measures the length of time the lamp is aglow, and a voltmeter that measures the voltage supplied by the battery. Again, let's assume that the wires have no resistance. Let's suppose the timer has no resistance either. That way, the wiring and the timer won't influence the outcome of our calculations, and we can be sure the voltmeter indicates the voltage across the lamp as well as the voltage provided by the battery.

Let E stand for the voltmeter reading (in volts), let I stand for the ammeter reading (in amperes), let R stand for resistance of the lamp (in ohms), and let t stand for the timer reading (in hours). Three formulas can be used to determine the energy, W (in watt-hours) consumed by the bulb:

$$W = EIt$$

$$W = E^2 t / R$$

$$W = I^2 Rt$$

You should use units of volts (V), amperes (A), ohms (Ω) and hours (h) for energy calculations to come out in watt-hours. If quantities are given in units other than these units, you must convert them to these units before you plug any numbers into the above formulas.

? **PROBLEM 3-11**

Suppose that in the circuit of Fig. 3-3, the switch is closed for 2 h and then opened again. While the switch is closed, the voltmeter indicates 12 V and the ammeter shows 1.5 A. What is the energy consumed, in watt-hours?

✔ **SOLUTION 3-11**

Use the first of the three formulas defined earlier. We're given the quantities in standard units, so there is no need to convert any of them. We can plug the numbers directly in. Here, $E = 12$, $I = 1.5$, and $t = 2$. Therefore:

$$W = EIt$$
$$= (12 \times 1.5 \times 2)\,\text{Wh}$$
$$= 36\,\text{Wh}$$

? **PROBLEM 3-12**

Imagine that we don't know the ammeter reading in the circuit of Fig. 3-3, but we know that the switch is closed for 30 minutes (30 min) and the voltmeter shows 6 V. The resistance of the lamp is known to be 6 Ω. How much energy is consumed by the lamp during this period of time?

✔ **SOLUTION 3-12**

First, convert the time to hours: 30 min = 0.5 h. Therefore, $t = 0.5$. Then note that $E = 6$ and $R = 6$. Use the second of the previously defined formulas for energy:

$$W = E^2 t / R$$
$$= (6 \times 6 \times 0.5/6)\,\text{Wh}$$
$$= 3\,\text{Wh}$$

? PROBLEM 3-13

Suppose we don't know the battery voltage, but we know that the lamp has a resistance of 10 Ω. The switch is closed for 90 min, and during this time the ammeter reads 400 mA. How much energy is consumed in this case?

✔ SOLUTION 3-13

First, convert the time to hours: 90 min = 1.5 h. Therefore, $t = 1.5$. Then convert the current to amperes: 400 mA = 0.4 A. Therefore, $I = 0.4$. Note that $R = 10$. Plug these numbers into the third formula for energy given above:

$$W = I^2 Rt$$
$$= (0.4 \times 0.4 \times 10 \times 1.5)\,\text{Wh}$$
$$= 2.4\,\text{Wh}$$

ENERGY VERSUS POWER

Energy is the equivalent of power multiplied by time, and power is the equivalent of energy per unit time. That's the essential difference between the two quantities. Energy is like distance traveled; power is like speed. In general, the power consumed rises and falls from moment to moment, but the overall energy consumption increases relentlessly with the passage of time.

Quiz

This is an "open book" quiz. You may refer to the text in this chapter. A good score is 8 correct answers. Answers are in Appendix 1.

1. Examine Fig. 3-3. Suppose that the original timer, which has no resistance, is replaced by another timer that has a resistance equal to that of the lamp. If nothing else in the circuit changes, how will the current through the lamp be different when the switch is closed, as compared with the current through the lamp in the original circuit when the switch is closed?

   (a) It will be the same.

(b)   It will be half as great.
(c)   It will be $^1/_4$ as great.
(d)   More information is needed to answer this question.

2.   Suppose a battery supplies 12 V to a DC circuit whose resistance is 144 kΩ. What is the DC power dissipated or used by this circuit?

(a)   1 W
(b)   1 mW
(c)   1 kW
(d)   More information is necessary to answer this question.

3.   Suppose an 8.2 MΩ resistor carries a direct current of 82 μA. What, approximately, is the DC voltage across this resistor?

(a)   670 V
(b)   1000 V
(c)   0.670 V
(d)   1 V

4.   Look at Fig. 3-3 again. Suppose that the original timer, which has no resistance, is replaced by another timer that has resistance equal to that of the lamp. If nothing else in the circuit changes, how will the voltage across the lamp (not the voltmeter reading) be different when the switch is closed, as compared with the voltage across the lamp in the original circuit when the switch is closed?

(a)   It will be the same.
(b)   It will be half as great.
(c)   It will be $^1/_4$ as great.
(d)   More information is needed to answer this question.

5.   Examine Fig. 3-3 once again. Suppose that the original timer, which has no resistance, is replaced by another timer that has a resistance equal to that of the lamp. If nothing else in the circuit changes, how will the voltmeter reading be different when the switch is closed, as compared with the voltmeter reading in the original circuit when the switch is closed?

(a)   It will be the same.
(b)   It will be half as great.
(c)   It will be $^1/_4$ as great.
(d)   More information is needed to answer this question.

6.   Examine Fig. 3-3 one more time. Suppose that the original timer, which has no resistance, is replaced by another timer that has resistance equal to that of the lamp. If nothing else in the circuit changes, how will the energy consumed by the lamp in 2 h be different when the switch is closed, as compared with the energy consumed by the lamp in 2 h in the original circuit when the switch is closed?

(a)   It will be the same.
(b)   It will be half as great.
(c)   It will be $^1/_4$ as great.
(d)   More information is needed to answer this question.

7.   How many ohms are in 47 kΩ?

(a)   0.0047 Ω
(b)   0.47 Ω
(c)   4700 Ω
(d)   47,000 Ω

8.   How many watts are in 2500 kWh?

(a)   2.5 W
(b)   0.0025 W
(c)   2,500,000 W
(d)   This question has no answer, because watts and kilowatt-hours do not express the same phenomenon.

9.   Suppose an electrical appliance draws 1500 W of power. What is the energy used by this appliance in 3 h?

(a)   4.5 Wh
(b)   500 Wh
(c)   4.5 kWh
(d)   More information is necessary to answer this question.

10.   If a device having a DC resistance of 3 Ω is supplied with a DC voltage of 3 V, how much energy does it consume?

(a)   3 Wh
(b)   333 mWh
(c)   9 Wh
(d)   It depends on how long the device is going to be in operation.

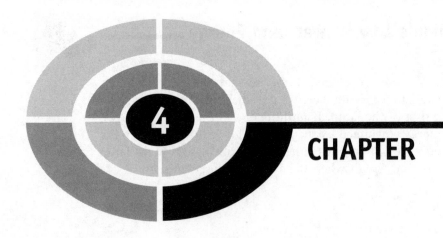

CHAPTER

Simple DC Circuits

A *DC circuit*, also called a *DC network*, is an interconnected set of components that operates from DC electricity. In this chapter, we'll examine a few DC circuits, and examine how the currents, voltages, and resistances behave in them.

Series Circuits

In a *series circuit*, all the components are connected end-to-end (Fig. 4-1). The current flows along a single branch, as water in a river without any tributaries or as through a simple garden hose.

VOLTAGE SOURCES IN SERIES

When DC voltage sources are connected in series, their voltages add up. Imagine *n* batteries in series (where *n* is a whole number), all connected "minus-to-plus." Let E_1, E_2, E_3, ..., E_n represent the voltages of the batteries, all expressed in volts. In theory, the total voltage, E, is the sum of

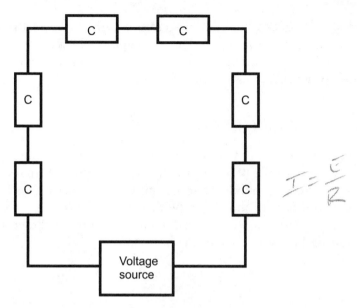

Fig. 4-1. In a series circuit, components (small boxes labeled C) are all connected end-to-end.

the voltages of the individual batteries, as shown in Fig. 4-2:

$$E = E_1 + E_2 + E_3 + \cdots + E_n$$

In real-world networks, this is an oversimplification. All the voltage sources have a little *internal resistance*. This makes the actual net voltage

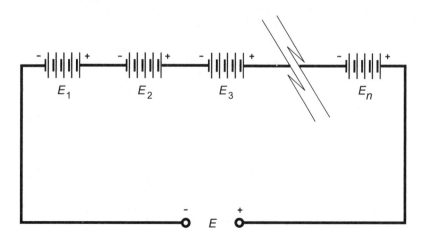

Fig. 4-2. When sources of DC voltage are connected in series, the total voltage is equal to the sum of the individual source voltages, assuming all the polarities are the same.

across a series combination of sources a little bit lower than the sum of the individual source voltages. This discrepancy increases as more and more current is demanded from the sources. Nevertheless, the above formula is good to remember as a basic principle.

POLARITY REVERSAL

Now imagine that not all the batteries are connected minus-to-plus. What if one or more of the batteries are "turned around"? In that case, the voltages of any "turned around" batteries must be subtracted from the total.

Suppose battery number 2 (with voltage E_2) in the circuit of Fig. 4-2 is connected with reversed polarity. This situation is shown in Fig. 4-3. The series-combination voltage E is now:

$$E = E_1 - E_2 + E_3 + \cdots + E_n$$

RESISTANCES IN SERIES

When resistors (components having resistance) are connected in series, their ohmic values add up. Imagine n resistors in series. Let R_1, R_2, R_3, ..., R_n represent their values, all expressed in ohms. The total resistance, R, is the sum of the resistances of the individual resistors, as shown in Fig. 4-4A:

$$R = R_1 + R_2 + R_3 + \cdots + R_n$$

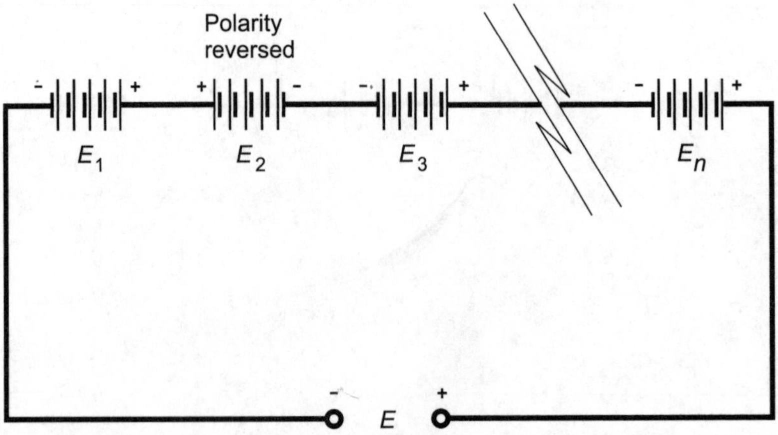

Fig. 4-3. A series combination of DC voltage sources in which the polarity of one source is reversed. This voltage subtracts from, rather than adds to, the total.

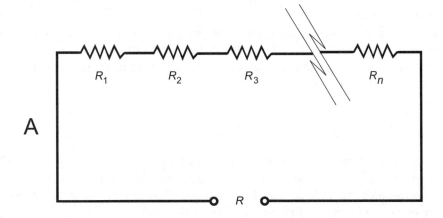

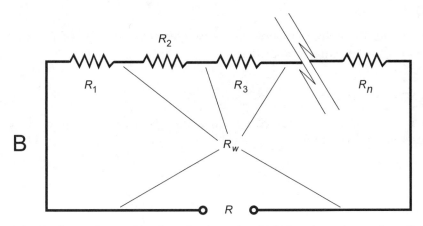

Fig. 4-4. At A, a series combination of resistors in which the wire resistance is negligible compared with the values of the resistors. At B, the wire resistance (R_w) is taken into account if the resistors have extremely small ohmic values.

This is a theoretical ideal, because it neglects the resistance of the wires that interconnect the resistors. In nearly all real-life circuits of this kind, the wire resistance can be considered equal to 0 Ω. However, there are occasional exceptions. If the ohmic values of the resistors themselves are so low that they are comparable with the resistance in an ordinary wire, this formula is an oversimplification. Then the wire itself, R_w, must be treated as another resistor in the formula, which becomes:

$$R = R_1 + R_2 + R_3 + \cdots + R_n + R_w$$

This is shown in Fig. 4-4B.

Polarity is irrelevant with plain resistors. In theory, resistors don't have poles. They can be turned around, and it makes no difference. The opposition to DC offered by a resistor in one direction is the same as the opposition to DC in the other direction.

CURRENT SOURCES IN "SERIES"

In a series DC circuit, there's only one branch in which current can flow. The current at any point is therefore the same as the current at any other point. It's meaningless to talk about current sources connected in series, because every component in a series circuit carries the same current, however large or small.

[?] **PROBLEM 4-1**

What is the DC voltage E between the two terminals in Fig. 4-5? Pay special attention to the battery polarities.

[✔] **SOLUTION 4-1**

The batteries are all connected with the same polarity. Therefore, their voltages add, and the total is:

$$E = (6 + 9 + 3 + 12)\,\mathrm{V}$$
$$= 30\,\mathrm{V}$$

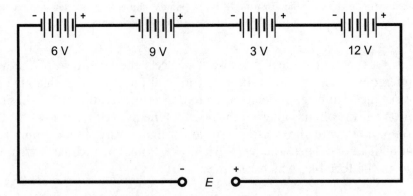

Fig. 4-5. Illustration for Problem 4-1.

? **PROBLEM 4-2**

What is the DC voltage E between the two terminals in Fig. 4-6? Pay special attention to the battery polarities.

✔ **SOLUTION 4-2**

Note that this circuit is the same as that in the previous problem, except that the two left-hand batteries have been "turned around." This means that their voltages subtract from, rather than add to, the total. The voltage at the terminals is therefore:

$$E = [(-6) + (-9) + 3 + 12]\,\mathrm{V}$$
$$= 0\,\mathrm{V}$$

The battery voltages collectively cancel each other out, so there is no voltage at the terminals.

? **PROBLEM 4-3**

What is the DC resistance R between the terminals in Fig. 4-7? Pay special attention to the resistance units. Express the answer in ohms. Assume the wire resistance is 0 Ω.

✔ **SOLUTION 4-3**

Convert all the values to ohms before doing any calculations. Note that 1.2 kΩ = 1200 Ω, and 1.8 kΩ = 1800 Ω. Therefore:

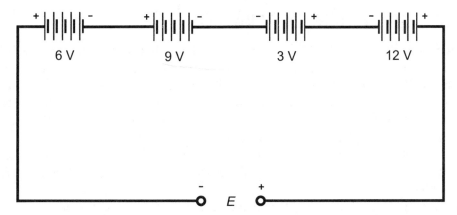

Fig. 4-6.   Illustration for Problem 4-2.

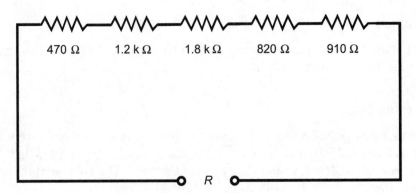

Fig. 4-7. Illustration for Problem 4-3.

$$R = (470 + 1200 + 1800 + 820 + 910)\ \Omega$$
$$= 5200\ \Omega$$

Parallel Circuits

In a *parallel circuit*, the components are arranged in such a way that they appear in a diagram like the rungs of a ladder (Fig. 4-8). Some current flows through every branch. The branch currents are not necessarily all the same. But every component receives the same voltage as every other; each individual component is directly connected across the voltage source.

VOLTAGE SOURCES IN PARALLEL

Normally, all the voltage sources in a parallel circuit have the same voltage. If the voltages differ, some sources drive current through others, and this is almost always a bad thing. In a parallel circuit, the poles of the voltage sources must all be connected plus-to-plus and minus-to-minus. Otherwise, there will be short-circuit loops containing pairs of sources in series. This wastes energy, and can even be dangerous.

The output voltage of a properly designed parallel combination of batteries is equal to the voltage of any single one of them. Batteries are combined in parallel to increase the current that the voltage source can deliver. The total deliverable current, I, of a parallel combination of n

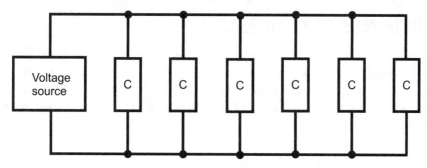

Fig. 4-8. In a parallel circuit, components (small boxes labeled C) are all connected across each other.

batteries is equal to the sum of the deliverable currents, I_1, I_2, I_3, ..., I_n, of each one (Fig. 4-9):

$$I = I_1 + I_2 + I_3 + \cdots + I_n$$

In the diagram, the arrows point in the direction of the theoretical currents, which, as we learned earlier, are considered to flow from positive poles to negative poles.

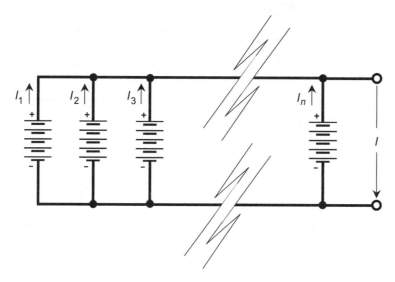

Fig. 4-9. When batteries having equal voltage are connected in parallel, the total deliverable current is equal to the sum of the individual deliverable currents, assuming all the polarities are the same.

RESISTANCES IN PARALLEL

When resistors are connected in parallel, their ohmic values combine in a rather complicated way. Consider n resistors in parallel. Let R_1, R_2, $R_3, \ldots, R_n$ represent their values, all expressed in ohms. If R is the total resistance in ohms, then its reciprocal, $1/R$, is given by the following formula (Fig. 4-10):

$$1/R = 1/R_1 + 1/R_2 + 1/R_3 + \cdots + 1/R_n$$

The total resistance is equal to the reciprocal of the sum of the reciprocals of the individual resistances. Therefore:

$$R = 1/(1/R_1 + 1/R_2 + 1/R_3 + \cdots + 1/R_n)$$

This assumes that the wire has no resistance. If the wire has a significant resistance (for example, if the resistors have extremely low ohmic values), then some of the wire resistance adds in series with each resistor, and some of the wire resistance adds in series with the whole combination. This can happen in so many different ways that there is no single formula that applies to all cases.

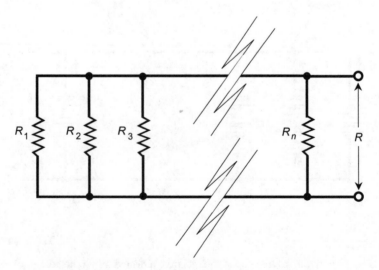

Fig. 4-10.   When DC resistances are connected in parallel, the composite resistance must be found using a rather cumbersome formula (see text).

THE SIEMENS

The reciprocal of resistance is known as *conductance*, and is expressed in a unit called the *siemens*. The word "siemens," which serves in the plural as well as the singular sense, is abbreviated by a non-italic, uppercase letter S. Conductance, as a variable, is symbolized in formulas by the uppercase italic letter G. If G is the conductance in siemens and R is the resistance in ohms, the following formulas hold true:

$$G = 1/R$$

and

$$R = 1/G$$

Conductance values in parallel add just like resistance values in series. If G_1, G_2, G_3, ..., G_n are the conductances, in siemens, of the individual resistors in a parallel combination, then the composite conductance G, also in siemens, is:

$$G = G_1 + G_2 + G_3 + \cdots + G_n$$

Some people find this formula easier to understand than the resistance formulas when dealing with parallel circuits. But if you try to calculate parallel resistances by converting them to conductances first, you must remember to convert the total conductance back to resistance at the end!

CURRENTS IN PARALLEL BRANCHES

Imagine a parallel circuit with n branches, connected across a DC voltage source, such as a battery (Fig. 4-11). Let I_1, I_2, I_3, ..., I_n represent the currents in each of the branches, all expressed in amperes. Then the total current, I, drawn from the battery, also in amperes, is the sum of the branch currents:

$$I = I_1 + I_2 + I_3 + \cdots + I_n$$

? **PROBLEM 4-4**

Suppose that 5 lantern batteries are connected in parallel. Each battery supplies 6 V. What is the voltage of the resulting combination?

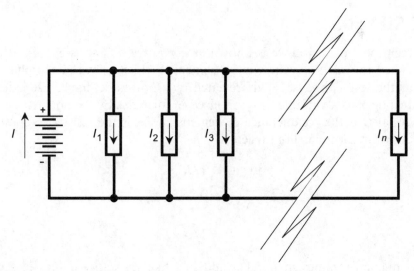

Fig. 4-11.   The total current drawn by a parallel circuit is equal to the sum of the currents in the individual branches.

✔ SOLUTION 4-4

The voltage of a parallel set of batteries, all having the same voltage, is equal to the voltage of any one considered by itself. Thus, the combination supplies 6 V.

? PROBLEM 4-5

Suppose, in the preceding problem, each battery can reliably provide up to 2 A of current. How much current can the combination reliably provide?

✔ SOLUTION 4-5

In theory, the combination can provide a current equal to the sum of the individual deliverable currents. This is $(6+6+6+6+6)$ A $= (5 \times 6)$ A $= 30$ A.

? PROBLEM 4-6

Examine Fig. 4-12. What is the current I_3?

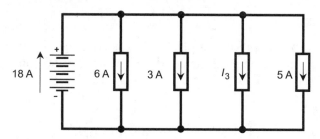

Fig. 4-12. Illustration for Problem 4-6.

✔ **SOLUTION 4-6**

This can be found by using the formula for the total current in a parallel combination, and using simple algebra to solve for I_3:

$$18 \text{ A} = (6 + 3 + I_3 + 5)\,\text{A}$$

Therefore

$$I_3 = 4\,\text{A}$$

? **PROBLEM 4-7**

What is the resistance R across the parallel combination of resistors shown in Fig. 4-13?

✔ **SOLUTION 4-7**

Use the formula for parallel resistances:

$$R = \left[1 \Big/ \left(\frac{1}{16} + \frac{1}{16} + \frac{1}{8} + \frac{1}{4} + \frac{1}{2} + 1 \right) \right] \Omega$$
$$= \left(\frac{1}{2} \right) \Omega = 0.5\,\Omega$$

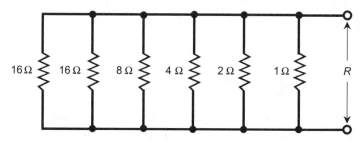

Fig. 4-13. Illustration for Problems 4-7 and 4-8.

> **[?] PROBLEM 4-8**

In the scenario of the preceding problem, convert the resistor values to conductances, and then calculate the conductance across the combination.

> **[✔] SOLUTION 4-8**

The conductance values, proceeding from left to right, are 1/16 S, 1/16 S, 1/8 S, 1/4 S, 1/2 S, and 1 S. Adding all these gives the conductance, G, across the combination:

$$G = \left(\frac{1}{16} + \frac{1}{16} + \frac{1}{8} + \frac{1}{4} + \frac{1}{2} + 1\right) S$$
$$= 2\,S$$

This is the reciprocal of the composite resistance in ohms.

Kirchhoff's Laws

Two of the most important DC network principles involve currents into and out of circuit points, and the sums of the voltages around closed loops. These rules are often called *Kirchhoff's First Law* and *Kirchhoff's Second Law*. They are also known as *Kirchhoff's Current Law* and *Kirchhoff's Voltage Law*, respectively.

THE LAW FOR CURRENT

The sum of the currents that flow into any point in a DC circuit is the same as the sum of the currents that flow out of that point. This is true no matter how many branches lead into or out of the point (Fig. 4-14). When making calculations, it's important to make sure all currents are specified in the same units.

This law arises from an almost trivial fact: electrical current can't appear from nowhere, and it can't vanish into nothingness. It's like water in a plumbing system: all the water that goes into any point in the pipes must come out of that point. (This is true even if the pipes leak!)

THE LAW FOR VOLTAGE

The sum of all the voltages, as you go around any closed loop in a DC circuit from some fixed point and return there from the opposite direction, and taking polarity into account, is always equal to 0 V (Fig. 4-15). As

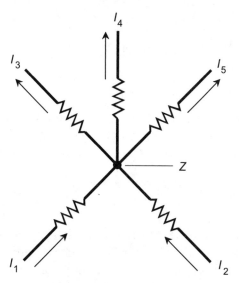

Fig. 4-14. Kirchhoff's Current Law. The total current going into point Z is the same as the total current going out.

with the law for current, it's important to make sure that all the voltages are specified in the same units.

Kirchhoff's Voltage Law arises from the fact that a potential difference can't exist between any circuit point and itself! A corollary to this is that a potential difference can't exist between any two points that are directly connected by a perfect conductor (that is, one that has theoretically zero resistance).

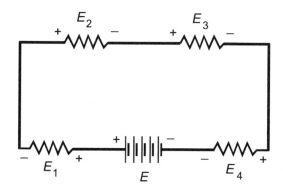

Fig. 4-15. Kirchhoff's Voltage Law. The sum of the voltages around the closed loop is equal to zero when polarities are taken into account.

? **PROBLEM 4-9**

Refer to Fig. 4-16. What is the current I_2?

✔ **SOLUTION 4-9**

Use Kirchhoff's Current Law. All the current values are given in amperes, so we don't have to convert any of the values to get them all into the same units. First, determine the total current I_{out} coming out of the branch point:

$$I_{out} = (4 + 5 + 7)\,A$$
$$= 16\,A$$

This means that the sum of the currents entering the branch point, I_{in}, must also be 16 A. We know two of the branch currents going into the point, so determining the unknown current, I_2, is a simple algebra problem:

$$16\,A = (6 + I_2 + 4)\,A$$

Therefore

$$I_2 = 6\,A$$

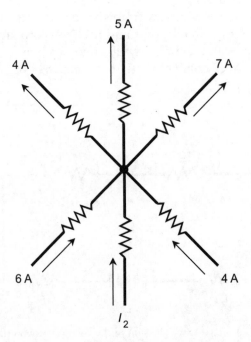

Fig. 4-16. Illustration for Problem 4-9.

? **PROBLEM 4-10**

Refer to Fig. 4-17. What is the battery voltage, E?

✔ **SOLUTION 4-10**

Use Kirchhoff's Voltage Law. The sum of all the voltages, taking polarity into account, must be equal to 0 V. Therefore:

$$E + (3 + 4 + 5 + 2)\,V = 0\,V$$
$$E + 14\,V = 0\,V$$
$$E = -14\,V$$

The answer comes out to be negative because we consider the potential differences across the resistors to be positive. The voltage across the battery has a polarity opposite the polarities of the potential differences across the resistors.

Quiz

This is an "open book" quiz. You may refer to the text in this chapter. A good score is 8 correct answers. Answers are in Appendix 1.

1. Refer to Fig. 4-18. The battery supplies 6 V. Suppose each of the lamps in the circuit has a resistance of 120 Ω. What is the resistance of the two lamps together, connected in parallel?

    (a) It can't be determined from this information.
    (b) 240 Ω
    (c) 120 Ω
    (d) 60 Ω

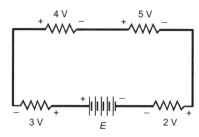

Fig. 4-17.   Illustration for Problem 4-10.

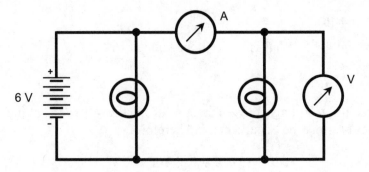

Fig. 4-18.   Illustration for Quiz Questions 1 through 5.

2.   Refer to Fig. 4-18. The battery supplies 6 V. Suppose each of the lamps in the circuit has a resistance of 120 Ω. What is the ammeter reading?

(a)   It can't be determined from this information.
(b)   50 mA
(c)   100 mA
(d)   200 mA

3.   Refer to Fig. 4-18. The battery supplies 6 V. Suppose each of the lamps in the circuit has a resistance of 120 Ω. What is the current drawn from the battery?

(a)   It can't be determined from this information.
(b)   50 mA
(c)   100 mA
(d)   200 mA

4.   Refer to Fig. 4-18. The battery supplies 6 V. Suppose each of the lamps in the circuit has a resistance of 120 Ω. What is the voltmeter reading?

(a)   It can't be determined from this information.
(b)   3 V
(c)   6 V
(d)   12 V

5. Refer to Fig. 4-18. The battery supplies 6 V. Suppose each of the lamps in the circuit has a resistance of 120 Ω. If the left-hand bulb blows out, leaving an open circuit in its place, what will happen to the ammeter reading?

    (a)   We cannot say from this information.
    (b)   It will stay the same.
    (c)   It will increase to twice its former reading.
    (d)   It will drop to half its former reading.

6. What is the voltage provided by 6 lantern batteries, each providing 6 V, connected in parallel (plus-to-plus and minus-to-minus)?

    (a)   1 V
    (b)   6 V
    (c)   36 V
    (d)   More information is needed to answer this question.

7. How many flashlight cells, each providing 1.5 V, must be connected in series (minus-to-plus) to provide 13.5 V for a radio receiver normally run from the electrical system of a car?

    (a)   4
    (b)   9
    (c)   12
    (d)   It is impossible to get 13.5 V by combining 1.5 V flashlight cells in series.

8. In order to determine the total current drawn by a parallel circuit, the currents in the individual branches

    (a)   must be combined according to an awkward formula, similar to the formula used to determine the total value of a parallel combination of resistors.
    (b)   can be added up, assuming all the branch currents are expressed in the same units.
    (c)   can be multiplied by each other, assuming all the branch currents are expressed in the same units.
    (d)   can be divided by the number of branches, assuming all the branch currents are expressed in the same units.

9. Suppose 10 batteries, each providing 6 V, are connected in series (minus-to-plus). If one of the batteries is turned around so that its polarity is reversed, the voltage across the entire series combination will

    (a)   increase by 12 V.
    (b)   increase by 6 V.
    (c)   decrease by 6 V.
    (d)   decrease by 12 V.

10. Suppose that six resistors, each having an ohmic value of 10 kΩ, are connected in series. What will happen to the total resistance of the combination if the polarity of one of the resistors is reversed?

    (a)   It will increase.
    (b)   It will decrease.
    (c)   A short circuit will result, with possible damage to the circuit.
    (d)   It will not change. Simple resistors do not have poles.

CHAPTER 5

Cells and Batteries

Electricity can be provided by chemical reactions. Another important source is sunlight. A third source is fuel that can be oxidized to obtain energy. In this chapter, we'll look at how DC is generated by these means.

Electrochemical Power

An *electrochemical cell* converts chemical energy (in this case, a form of *potential energy*) into electrical energy. When two or more such cells are connected in series, the result is an *electrochemical battery*. Electrochemical cells and batteries are used in portable electronic equipment, in communications satellites, and as sources of emergency power.

ELECTROCHEMICAL ENERGY

Figure 5-1 shows an example of a *lead-acid cell*. An electrode of lead and an electrode of lead dioxide, immersed in a sulfuric-acid solution called the *electrolyte*, acquire a potential difference. This voltage can drive current through a load. The *maximum deliverable current* depends on the mass and

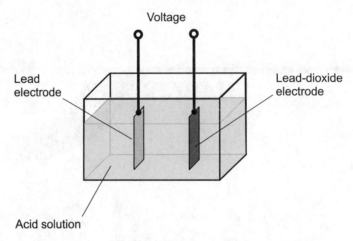

Fig. 5-1. A lead-acid cell.

volume of the cell. In a battery made from lead-acid cells, the voltage depends on the number of cells.

If a lead-acid cell is connected to a load for a long time, the current gradually decreases, and the electrodes become coated. The nature of the acid changes, too. Eventually, all the chemical energy contained in the acid is converted into electrical energy. Then the current drops to zero, and a potential difference no longer exists between the two electrodes.

PRIMARY AND SECONDARY CELLS

Some electrochemical cells, once their chemical energy has all been changed to electricity and used up, must be discarded. This type of cell is called a *primary cell*. Other kinds of cells, like the lead-acid unit described above, can get their chemical energy back again by recharging. Such a component is called a *secondary cell*.

Most primary cells contain a chemical paste along with metal electrodes. They go by names such as *dry cell*, *zinc-carbon cell*, or *alkaline cell*. They are commonly found in supermarkets and department stores. Some secondary cells can also be found in stores. These cost more than the ordinary dry cells, and a charging unit also costs a few dollars. But these rechargeable cells can be used hundreds of times, and can pay for themselves and the charger, several times over.

An *automotive battery* is made from secondary cells connected in series. These cells recharge from the vehicle's alternator or from an outside charging unit. This battery has cells like the one shown in Fig. 5-1. It is dangerous to short-circuit the terminals of such a battery, or even to subject it to an exceptionally heavy load (demand excessive current from it), because the acid can boil out. In fact, it is unwise to short-circuit any cell or battery, because it might explode or cause a fire.

STANDARD CELL

Most cells produce a potential difference of 1.0 V to 1.8 V between their positive and negative electrodes. Some types of cells generate predictable and precise voltages. These are called *standard cells*. An example is the *Weston standard cell*, which produces 1.018 V at room temperature. It has an electrolyte solution of cadmium sulfate. The positive electrode is made of mercury sulfate, and the negative electrode is made of mercury and cadmium (Fig. 5-2).

When a Weston standard cell is properly constructed and used at room temperature, its voltage is always the same. This allows it to be employed as a voltage standard.

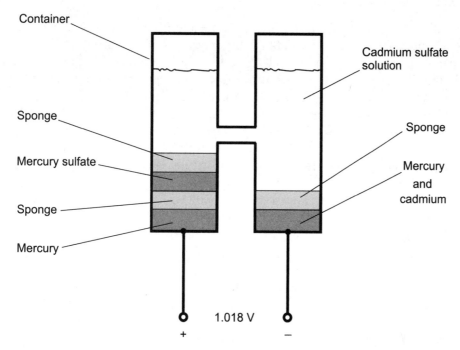

Fig. 5-2.   A Weston standard cell.

STORAGE CAPACITY

Any cell or battery has a certain amount of electrical energy that can be specified in watt hours (Wh) or kilowatt hours (kWh). Often, it is given in terms of the mathematical *integral* of deliverable current with respect to time, in units of ampere hours (Ah). The energy capacity in watt hours is the ampere-hour capacity multiplied by the battery voltage.

A battery with a rating of 20 Ah can provide 20 A for 1 h, or 1 A for 20 h, or 100 mA for 200 h. There exists an infinite number of current/time combinations, and almost any of them (except for the extremes) can be put to use in real life. The extreme situations are the *shelf life* and the *maximum deliverable current*. Shelf life is the length of time the battery will remain usable if it is never connected to a load; this can be years. The maximum deliverable current is the highest current a battery can drive through a load without the voltage dropping significantly because of the battery's internal resistance, and without creating a fire or explosion hazard.

POLARITY

A cell or battery always has a positive pole and a negative pole. Theoretical (conventional) current, as defined by physicists, flows from the positive pole to the negative pole of a cell or battery that is connected in an electrical circuit. The electrons, however, move from the negative to the positive.

In some electrical systems, DC polarity doesn't matter. In a lantern, the battery can be connected in either direction, and the lamp will light up, provided its electrical contact is properly maintained. If you turn all the cells around in a flashlight, and if the bulb still makes contact and all the cells are touching, the device will work in the same way as it does when the cells are installed "correctly." If you turn only one of the cells around in a two-cell flashlight, however, the bulb won't illuminate because the voltages from the cells cancel each other out.

In most electronic devices, battery polarity is critical. A cell or battery must be connected in such a circuit "the right way." Otherwise the device or system won't work, and damage can occur to some of its components. Portable electronic devices in which the battery polarity matters include radio receivers and transmitters, remote control boxes, cell phones, pagers, tape players, compact disc (CD) and digital video disc (DVD) players, electric clocks, and notebook computers. (This is by no means a complete list, so beware!)

DISCHARGE CURVES

When an *ideal cell* or *ideal battery* is used, it delivers a constant current for awhile, and then the current starts to decrease. Some types of cells and batteries approach this ideal behavior, exhibiting a *flat discharge curve* (Fig. 5-3A). Others have current that decreases gradually from the beginning of use; this is a *declining discharge curve* (Fig. 5-3B).

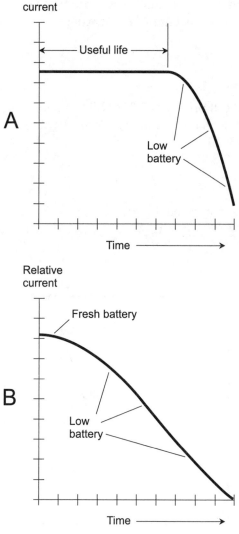

Fig. 5-3.   At A, a flat discharge curve. At B, a declining discharge curve.

When the current that a battery can provide has decreased to about half of its initial value, the cell or battery is said to be "weak" or "low." At this time, it should be replaced. If it is allowed to run down until the current drops to nearly zero, the cell or battery is said to be "dead," although, in the case of a rechargeable unit, a better term is "depleted."

? PROBLEM 5-1

Suppose a Weston standard cell is connected to a voltmeter, and the meter produces a reading such as that shown in Fig. 5-4. From this, we can tell that the meter is not accurate. Approximately how far off is the meter reading?

✔ SOLUTION 5-1

A Weston standard cell produces 1.018 V. The meter indicates approximately 1.5 V. This means the meter reading is almost 0.5 V too high. This represents an error of nearly 50%.

? PROBLEM 5-2

Suppose the meter portrayed in Fig. 5-4 always reads 50% too high. If a flashlight cell that provides 1.5 V is connected to the meter, what should we expect?

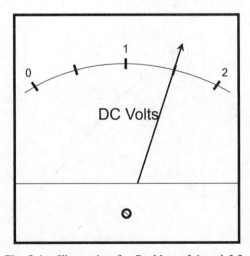

Fig. 5-4. Illustration for Problems 5-1 and 5-2.

✔ **SOLUTION 5-2**

We should expect that the meter needle will "hit the pin." A reading 50% above 1.5 V is equal to 1.5×1.5 V, or 2.25 V. This is beyond the full-scale range of the meter.

? **PROBLEM 5-3**

How might the meter portrayed in the previous two problems be made accurate?

✔ **SOLUTION 5-3**

The meter can be connected in series with a potentiometer having an extremely large value (Fig. 5-5). Then a Weston standard cell can be connected across the combination. The potentiometer should be adjusted until the meter reading is slightly more than 1 V. This will correct the error over the entire range of the meter, provided the circuit within the dashed box (and not just the meter itself) is used to measure DC voltages between 0 V and 2 V.

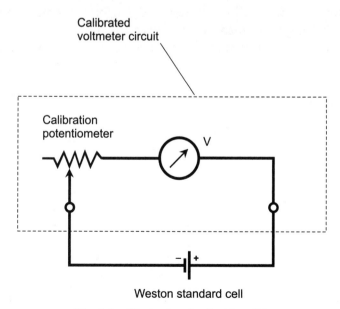

Fig. 5-5. Illustration for Problem 5-3.

Common Cells and Batteries

The cells sold in stores, and used in convenience items like flashlights and transistor radios, provide approximately 1.5 V, and are available in sizes AAA (very small), AA (small), C (medium), and D (large). Batteries made from combinations of these cells are usually rated at 6 V (4 cells in series) or 9 V (6 cells in series).

ZINC-CARBON

Zinc-carbon cells have a fairly long shelf life. The zinc forms the outer case and is the negative electrode. A carbon rod serves as the positive electrode. The electrolyte is a paste of manganese dioxide and carbon. Zinc-carbon cells are inexpensive and are usable at moderate temperatures, and in applications where the current drain is from moderate to high. They do not work well in extremely cold environments.

ALKALINE

Alkaline cells have granular zinc for the negative electrode, potassium hydroxide as the electrolyte, and a substance called a *polarizer* as the positive electrode. An alkaline cell can work at lower temperatures than a zinc-carbon cell. It lasts a long time in low-current electronic devices, and is therefore preferred for use in transistor radios, calculators, and portable cassette players. Its shelf life is much longer than that of a zinc-carbon cell.

TRANSISTOR

Transistor batteries are small, box-shaped batteries with clip-on connectors on top. They provide 9 V and consist of 6 zinc-carbon or alkaline cells in series. Transistor batteries are used in very-low-current electronic devices that are operated on an intermittent basis, such as radio-controlled garage-door openers, television and stereo remote-control boxes, smoke detectors, and electronic calculators.

LANTERN

Lantern batteries are rather massive and can deliver a fair amount of current. One type has spring contacts on the top. The other type has thumbscrew terminals. Besides keeping an incandescent bulb lit for awhile, these batteries, usually rated at 6 V and consisting of 4 zinc-carbon or alkaline cells connected in series, can provide enough energy to operate a low-power two-way communications radio.

SILVER-OXIDE

Silver-oxide cells are usually made into button-like shape. For this reason, they are sometimes called *button cells*, although some other types of cells also have this shape. They fit inside wristwatches, subminiature calculators, and small cameras. They come in various sizes and thicknesses, supply 1.5 V, and offer excellent energy storage for the weight. They are known to have a flat discharge curve. Silver-oxide cells can be stacked to make batteries about the size of an AA cylindrical cell.

MERCURY

Mercury cells, also called *mercuric oxide cells*, have advantages similar to the silver-oxide cells. They are manufactured in the same button-like shape. The main difference, often not of significance, is a somewhat lower voltage per cell: approximately 1.35 V.

There has been a decrease in the popularity of mercury cells and batteries in recent years, because mercury is toxic. When discarding mercury cells, special precautions must be taken, and these precautions are mandated by laws that vary from one locale to another. If you have cells or batteries that you suspect contain mercury, call your local trash-removal department, and get instructions on how to dispose of the cells or batteries.

LITHIUM

Lithium cells supply 1.5 V to 3.5 V, depending on the chemistry used. These cells, like their silver-oxide cousins, can be stacked to make batteries. Lithium cells and batteries have superior shelf life, and they can last for years in very-low-current applications. They provide excellent energy capacity per unit volume.

LEAD-ACID

Lead-acid cells and batteries have an electrolyte of sulfuric acid, along with a lead electrode (negative) and a lead-dioxide electrode (positive).

Some lead-acid batteries contain an electrolyte that is thickened into a paste. These are sometimes used in consumer devices that require moderate current, such as notebook computers, handheld computers, and portable video disc players. They are also used in uninterruptible power supplies (UPSs) for desktop computer systems.

NICKEL

Nickel-based cells and batteries are available in various configurations. *Cylindrical cells* look like size AAA, AA, C, or D zinc-carbon or alkaline cells. Nickel-based button cells are used in cameras, watches, memory backup applications, and other places where miniaturization is important. *Flooded cells* are used in heavy-duty applications, and can have a storage capacity of as much as 1000 Ah. They have a box-like appearance. *Spacecraft cells* are made in super-strong, thermally insulated packages that can withstand extraterrestrial temperatures and pressures.

Nickel-cadmium (NICAD) batteries are available in box-shaped packages that can be plugged into any equipment to form a part of the case for a device. An example is the battery pack for a handheld radio transceiver. This type of battery should never be left connected to a load after it has discharged. This can cause the polarity of one or more cells to reverse. Once this happens, the battery will not accept a recharge and will no longer be usable. When a NICAD cell or battery has discharged almost all the way, it should be recharged as soon as possible.

Nickel-metal-hydride (NiMH) cells and batteries can directly replace NICAD units in most applications. These batteries are in fact preferred, because they do not contain toxic cadmium. Some engineers believe that NiMH batteries also exhibit better behavior when repeatedly discharged and recharged.

[?] **PROBLEM 5-4**

Suppose you want to make a 9 V battery by stacking silver-oxide cells. How many cells will it take? How can you be sure the polarity is correct?

[✔] **SOLUTION 5-4**

A single silver-oxide cell produces 1.5 V. Therefore, 6 such cells are required. Connected in series, they produce 6×1.5 V $= 9$ V. It is important that all

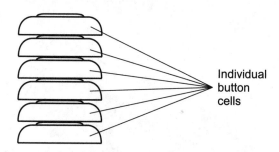

Fig. 5-6. Illustration for Problem 5-4.

the individual button cells be stacked so that they're oriented in the same direction (as shown in Fig. 5-6). It is also important to pay attention to the polarity when using the battery. One face of each button cell should be labeled with either a plus sign (+) or a minus sign (−). If this is not the case, the battery should be tested for polarity with an inexpensive multimeter, available at most hardware or electronics stores.

? **PROBLEM 5-5**

Suppose you can't find silver-oxide button cells, but only mercury button cells are available. How many of these will it take to produce 9 V?

✔ **SOLUTION 5-5**

In order to figure this out, first divide 9 V by 1.35 V, which is the voltage provided by a single mercury cell. This gives:

$$(9V)/(1.35V) = 6.67$$

Obviously we can't stack a fractional number of cells. (Even if we could cut the cells apart, which we would definitely not want to do with cells containing mercury, it would not change their voltage!) We can therefore do either of two things: settle for 6×1.35 V $= 8.1$ V, or deal with 7×1.35 V, or 9.45 V. In most applications, either approach will produce satisfactory results.

Photovoltaics

A *photovoltaic (PV) cell* is a *semiconductor device* that converts visible light, infrared radiation, or ultraviolet radiation directly into electricity. This type of cell is completely different from an electrochemical cell. It's also commonly

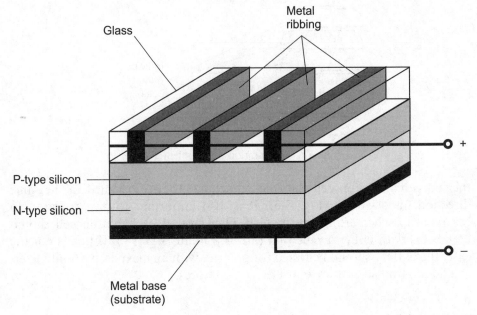

Fig. 5-7. Construction of a silicon photovoltaic cell.

known as a *solar cell*. Photovoltaic cells can be combined into *photovoltaic batteries* to obtain considerable voltage and power output.

OPERATION

The basic structure of a silicon PV cell is shown in Fig. 5-7. It is made of two types of specially treated material, called the *P type silicon* and the *N type silicon*. The top of the assembly is transparent so that light can fall directly on the P-type material. Metal *ribbing*, which forms the positive electrode, is interconnected by means of tiny wires. The negative electrode is a metal backing called the *substrate*, placed in contact with the N type silicon.

When visible light, infrared radiation, or ultraviolet radiation strikes the boundary surface between the P type and the N type silicon (known as the *P–N junction*), a potential difference is produced. The intensity of the current from the PV cell, under constant load conditions, varies in direct proportion to the brightness of the light striking the device, up to a certain point. Beyond that point, the increase becomes more gradual, and it finally levels off at a maximum current called the *saturation current*. The ratio of the available output power to the light power striking a photovoltaic cell is called the *conversion efficiency* of the cell.

Silicon solar cells produce about 0.6 V DC when exposed to light of sufficient brilliance. If there is low current demand, it doesn't take very bright light to produce the full output voltage. As the current demand increases, brighter and brighter light is needed to produce the full output voltage.

There is a maximum limit to the current that can be provided from a PV cell, no matter how bright the light is. This limit depends on the surface area of the cell. If more current is required than a single PV cell can deliver, then two or more cells can be connected in parallel. If more voltage is needed than a single cell can produce, then two or more cells can be connected in series.

SOLAR PANELS

Photovoltaic cells can be combined in series–parallel to make a *solar panel*. Combinations that contain a large number of PV cells are called *PV matrixes* or *PV arrays*. For example, a 50 parallel-connected set of 24 series-connected cells provide approximately 13 V with substantial deliverable current. Some solar panels are extremely large, covering hundreds of square meters.

Although high voltages (such as 500 V) can theoretically be obtained by connecting many photovoltaic cells in series, this is not practical because the internal resistances of the cells add up, limiting the current output and causing voltage-regulation problems. If high voltage is needed from a solar panel, a device called a *power inverter* can be used along with a high-capacity rechargeable battery, such as a lead-acid automotive type. Power inverters are available at large department stores, home supply stores, and hardware stores. The solar panel keeps the battery charged; the battery delivers high current on demand to the power inverter. Such a system provides 117 V AC from a 12 V to 14 V DC source.

SOLAR-ELECTRIC POWER

An independent solar/battery power system is called a *stand-alone solar electric energy system*. It uses a large solar panel, a large-capacity lead-acid or a nickel-based battery, a power inverter to convert the low-voltage DC into utility AC, and a sophisticated charging circuit. Such a system is best suited to environments where there is sunshine a high percentage of the time.

Solar cells, either alone or supplemented with rechargeable batteries, can be incorporated into an *interactive solar electric energy system*. This requires a special arrangement with the electric utility company. When the solar power system can't provide for the needs of the household or business all by itself, the utility company can take up the slack. Conversely, when the solar power system supplies more than enough energy for the needs of the home, the utility company can buy the excess.

? PROBLEM 5-6

Suppose you have access to an unlimited number of identical silicon PV cells. Each cell supplies 0.6 V DC at up to 50 mA in direct sunlight. You want to produce a power source sufficient to operate a device that requires 12 V DC at 1 A. What is the smallest possible PV array that will accomplish this, assuming that you have an environment where bright, direct sunlight shines when you need to use the device?

✔ SOLUTION 5-6

First, note that you need a source that will supply 12 V DC, which means that you must connect sets of 20 PV cells in series, because $20 \times 0.6\,V = 12\,V$. When two or more identical PV cells are connected in series, the maximum deliverable current of the combination is the same as that of any one of the cells. In this case, that is 50 mA.

In order to get 1 A of current, you'll have to combine 20 of the series PV sets in parallel. This is because $20 \times 50\,mA = 1000\,mA = 1\,A$. The resulting series-parallel array contains 400 PV cells, and is a 20-by-20 matrix.

? PROBLEM 5-7

In the above situation, suppose you build sets of 20 PV cells in parallel, and then combine 20 of these parallel sets in series. Will this work as well as the arrangement described in Solution 5-6, or not?

✔ SOLUTION 5-7

This will work just as well as the arrangement described in Solution 5-6. Instead of having 20 sources of 12 V, each capable of delivering 50 mA and all connected in parallel, you will have 20 sources of 0.6 V, each capable of delivering 1 A and all connected in series. The two matrixes, while different in their interconnection geometry, produce the same results in practice.

Fuel Cells

In the late 1900s, a new type of electrochemical power device emerged that is believed by many scientists and engineers to hold promise as an alternative energy source: the *fuel cell*.

HYDROGEN FUEL

The most talked-about fuel cell during the early years of research and development became known as the *hydrogen fuel cell*. As its name implies, it derives electricity from hydrogen. The hydrogen combines with oxygen (that is, it *oxidizes*) to form energy and water. There is no pollution and there are no toxic by-products. When a hydrogen fuel cell "runs out of juice," all it needs is a new supply of hydrogen, because its oxygen is derived from the atmosphere.

Instead of combusting, the hydrogen in a fuel cell oxidizes in a more controlled fashion, and at a much lower temperature. There are several schemes for making this happen. The *proton exchange membrane (PEM) fuel cell* is one of the most widely used fuel cells. A PEM hydrogen fuel cell generates approximately 0.7 V DC. In order to obtain higher voltages, individual cells are connected in series. A series-connected set of fuel cells is technically a battery, but the more often-used term is *stack*.

Fuel-cell stacks are available in various sizes. A stack about the size and weight of an airline suitcase filled with books can power a subcompact electric car. Smaller cells, called *micro fuel cells*, can provide DC to run devices that have historically operated from conventional cells and batteries. These include portable radios, lanterns, and notebook computers.

OTHER FUELS

Hydrogen is not the only chemical that can be used to make a fuel cell. Almost anything that combines with oxygen to form energy has been considered.

Methanol, which is a form of alcohol, has the advantage of being easier to transport and store than hydrogen, because it exists as a liquid at room temperature. *Propane* is another chemical that has been used for powering fuel cells. This is the substance that is stored in liquid form in tanks for barbecue grills and some rural home heating systems. *Methane*, also known as natural gas, has been used as well.

Some scientists and engineers object the use of these fuels because they, especially propane and methane, closely resemble fuels that are already commonplace, and on which society has developed the sort of dependence that purists would like to get away from. In addition, they are derived from so-called *fossil fuel* sources, the supplies of which, however great at the moment, are nevertheless finite.

A PROMISING TECHNOLOGY

Until now, fuel cells have not replaced the conventional electrochemical cells and batteries. Cost is the main reason. Hydrogen is the most abundant and the simplest chemical element in the universe, and it does not produce any toxic by-products. This would at first seem to make it the ideal choice for use in fuel cells. But storage and transport of hydrogen has proven to be difficult and expensive. This is especially true for fuel cells and stacks intended for operating systems that aren't fixed to permanent pipelines.

An interesting scenario, suggested by one of my physics teachers all the way back in the 1970s, is the piping of hydrogen gas through the standard utility lines designed to carry methane. Some modification of the existing system would be required in order to safely handle hydrogen, which is lighter than methane and can escape through small cracks and openings more easily. But hydrogen, if obtained at a reasonable cost and in abundance, could be used to power large fuel-cell stacks in common households and businesses. The DC from such a stack could be converted to utility AC by means of the same types of power inverters now used with PV energy systems. The entire home power system would be about the same size as a gas furnace.

Quiz

This is an "open book" quiz. You may refer to the text in this chapter. A good score is 8 correct answers. Answers are in Appendix 1.

1. Suppose you need a rechargeable battery that can provide 12 V DC at 7.5 A for 10 h, both day and night. Which of the following is the right choice?

    (a) A zinc-carbon battery.
    (b) A transistor battery.
    (c) A silicon PV battery.
    (d) An automotive lead-acid battery.

2.   The maximum deliverable current from a lead-acid battery depends on

(a)   the mass and volume of the battery.
(b)   the number of cells connected in series.
(c)   the intensity of visible light striking the cells.
(d)   the output voltage.

3.   A 10-by-10 series-parallel array of silicon PV cells, exposed to direct sunlight, can be expected to produce

(a)   0.06 V.
(b)   0.6 V.
(c)   6 V.
(d)   60 V.

4.   A stand-alone solar electric energy system requires

(a)   a storage battery of some sort.
(b)   a fuel cell.
(c)   a connection to the electric utility.
(d)   All of the above.

5.   If a NICAD battery is allowed to discharge completely, to the extent that the polarity of one or more of its cells reverses, what should you do?

(a)   Recharge it with a high-voltage charger.
(b)   Turn it around and connect it back into the circuit.
(c)   Connect it in series with another battery.
(d)   Replace it.

6.   Examine Fig. 5-8. This is a set of four silicon PV cells connected in parallel. What is wrong with this circuit?

(a)   The polarities should all agree, so the cells are connected minus-to-minus and plus-to-plus.
(b)   The polarities should all alternate, so the cells are connected minus-to-plus and plus-to-minus.
(c)   Silicon PV cells should never be connected in parallel.
(d)   Nothing is wrong with this circuit.

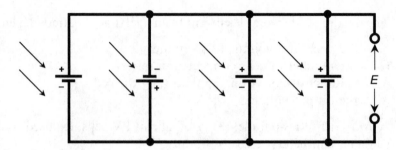

Fig. 5-8. Illustration for Quiz Questions 6 and 7.

7. In the circuit of Fig. 5-8, suppose the second-from-the-left PV cell is removed. What will be the DC voltage E when the array is exposed to bright sunlight and there is no load connected?

    (a) There is no way to tell without more information
    (b) Approximately 1.8 V
    (c) Approximately 0.6 V
    (d) Approximately 0.2 V

8. A rechargeable cell is also generally known as
    (a) a standard cell
    (b) a lead-acid cell
    (c) a secondary cell
    (d) a fuel cell

9. Fill in the blank to make the following sentence true: "When a _____ cell has been almost completely used up, it can be recharged and used again."

    (a) silicon
    (b) NICAD
    (c) solar
    (d) zinc-carbon

10. The length of time a battery will keep its useful charge, if it is allowed to sit around and is never used to provide power to anything, is known as its

    (a) maximum deliverable time.
    (b) shelf life.
    (c) discharge period.
    (d) depletion curve.

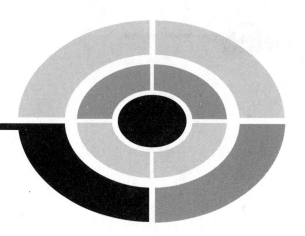

Test: Part One

Do not refer to the text when taking this test. You may draw diagrams or use a calculator if necessary. A good score is at least 30 answers (75% or more) correct. Answers are in the back of the book. It's best to have a friend check your score the first time, so you won't memorize the answers if you want to take the test again.

1. High voltage is, in general, more dangerous than low voltage because

    (a)   high voltage is more capable than low voltage of driving a lethal current through your body, if all other factors are held constant.

    (b)   the resistance of your body decreases as the voltage decreases, if all other factors are held constant.

    (c)   the conductance of your body decreases as the voltage increases, if all other factors are held constant.

    (d)   the electrical charge on your body diminishes as the voltage decreases, if all other factors are held constant.

    (e)   Stop! The premise is false. High voltage isn't any more dangerous than low voltage, whether other factors are held constant, or not.

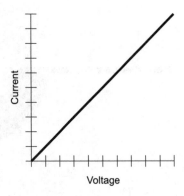

Fig. Test1-1. Illustration for Part One Test Questions 3 and 4.

2. A voltage of 200 μV is the equivalent of

   (a)   200,000 mV.
   (b)   20,000 mV.
   (c)   2000 mV.
   (d)   20 mV.
   (e)   None of the above

3. Figure Test1-1 is a graph of

   (a)   the current across a voltage as a function of the potential difference through it.
   (b)   the resistance as a function of the current across a component.
   (c)   the current through a component as a function of the voltage across it.
   (d)   All three of the above (a), (b), and (c) are true
   (e)   None of the above (a), (b), or (c) is true

4. From Fig. Test1-1, it is apparent that for this particular component,

   (a)   the resistance does not change as the current through it changes, at least within the range shown.
   (b)   the resistance does not change as the voltage across it changes, at least within the range shown.
   (c)   the current/voltage function is a straight line, at least within the range shown.
   (d)   All three of the above (a), (b), and (c) are true
   (e)   None of the above (a), (b), or (c) is true

5. Suppose there are 5 resistors, each with a value of 820 Ω, are connected in series. What is the net resistance of the combination?

   (a) 164 Ω
   (b) 820 Ω
   (c) 4.1 kΩ
   (d) 8.2 kΩ
   (e) More information is needed to answer this question.

6. Fill in the blank in the following sentence to make it true: "The cells you find in stores, such as zinc-carbon or alkaline size D flashlight cells, can be placed in series to get _____ than is possible with only one cell."

   (a) more current
   (b) more resistance
   (c) more voltage
   (d) longer working life
   (e) a better discharge curve

7. Suppose you are drawing a schematic diagram, and you have to make two lines (representing wires) cross on the page. You want to indicate that they are connected. You can do this by

   (a) drawing the two lines and letting them cross, but not putting a dot at the point of intersection.
   (b) drawing the two lines and letting them cross, putting a solid black dot at the point of intersection.
   (c) drawing the two lines and letting them cross, but making a little "jog" in one of the lines at the point of intersection.
   (d) drawing the two lines and letting them cross, making little "jogs" in both of the lines at the point of intersection.
   (e) finding some way to draw the diagram so you don't have to make the lines cross.

8. How many elementary charge units (ECU) are there in a volt?

   (a) 0.001
   (b) 1,000
   (c) 1,000,000
   (d) 6,240,000,000,000,000,000
   (e) This question cannot be answered, because charge and voltage are two different phenomena.

9. The current in an electrical conductor is considered by physicists to flow from the positive pole to the negative pole. Nevertheless, in a conventional conductor, the particles responsible for the current are

    (a) negatively charged ohms.
    (b) negatively charged neutrons.
    (c) negatively charged protons.
    (d) negatively charged electrons.
    (e) negatively charged volts.

10. Two objects will tend to be attracted toward each other if

    (a) they are both electrically neutral.
    (b) they are both electrically charged, one positively and the other negatively.
    (c) they are both electrically charged, and both charges are positive.
    (d) they are both electrically charged, and both charges are negative.
    (e) Any of the above conditions exists

11. A resistance of $20\,M\Omega$ is the equivalent of

    (a) $0.00002\,\Omega$
    (b) $0.02\,\Omega$
    (c) $20,000\,\Omega$
    (d) $20,000,000\,\Omega$
    (e) None of the above

12. Suppose that 5 resistors, each with a value of $820\,\Omega$, are connected in parallel. What is the net resistance of the combination?

    (a) $164\,\Omega$
    (b) $820\,\Omega$
    (c) $4.1\,k\Omega$
    (d) $8.2\,k\Omega$
    (e) More information is needed to answer this question.

13. Suppose that 6 flashlight cells, each producing $1.5\,V$, are connected in series in plus-to-minus fashion (that is, so that any two terminals that touch each other have opposite polarity). What is the net voltage of the combination?

    (a) $0\,V$
    (b) $1.5\,V$

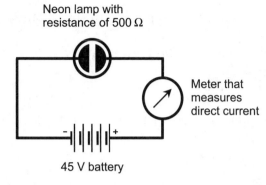

Neon lamp with
resistance of 500 Ω

Meter that
measures
direct current

− + 45 V battery

Fig. Test1-2.   Illustration for Part One Test Questions 14 through 18.

(c)   4.5 V
(d)   9 V
(e)   More information is necessary to answer this question.

14.   In the circuit shown by Fig. Test1-2, suppose the battery consists of
       1.5 V zinc-carbon cells in series. How many cells are there in this
       battery?

(a)   5
(b)   10
(c)   15
(d)   30
(e)   It is impossible to tell without more information.

15.   In the circuit shown by Fig. Test1-2, what does the meter read,
       assuming it is calibrated properly and is of the correct range to
       give a meaningful reading of the DC in this circuit?

(a)   90 μA
(b)   90 mA
(c)   11.1 mA
(d)   11.1 A
(e)   It is impossible to tell without more information.

16.   Suppose several more neon lamps, all identical to the one there
       now, are connected in parallel with the battery in the circuit of
       Fig. Test1-2. What will happen to the current drawn from the
       battery?

(a)   It will increase in direct proportion to the number of lamps.
(b)   It will decrease in direct proportion to the number of lamps.

(c) It will increase in proportion to the square of the number of lamps.

(d) It will decrease in proportion to the square of the number of lamps.

(e) It will not change.

17. How much power is dissipated by the neon lamp in the circuit of Fig. Test1-2?

(a) 4.05 W
(b) 11.1 W
(c) 90 mW
(d) 247 mW
(e) It is impossible to tell without more information.

18. How much energy is dissipated by the neon lamp in the circuit of Fig. Test1-2?

(a) 4.05 Wh
(b) 11.1 Wh
(c) 90 mWh
(d) 247 mWh
(e) It is impossible to tell without more information.

19. A button cell would most likely be used as the main source of power for

(a) a wristwatch.
(b) a lantern.
(c) a portable compact disc player.
(d) a 117-V utility circuit.
(e) Any of the above

20. Suppose you let a 60 W bulb stay aglow continuously. Energy in your location costs 10 cents per kilowatt hour. How much does it cost to leave this bulb on for 10 days and 10 nights?

(a) $0.15
(b) $1.44
(c) $14.40
(d) $86.40
(e) This question can't be answered unless we know the utility voltage.

21. Suppose that 16 resistors, each with a value of 820 Ω, are connected in a 4-by-4 series-parallel matrix. What is the net resistance of the combination?

   (a) 164 Ω
   (b) 820 Ω
   (c) 4.1 kΩ
   (d) 8.2 kΩ
   (e) More information is needed to answer this question.

22. Suppose a single solar cell is capable of supplying exactly 81 mW of DC power when exposed to direct sunlight. If 3 of these cells are connected in series, the resulting set of solar cells can theoretically provide

   (a) 9 mW of DC power in direct sunlight.
   (b) 27 mW of DC power in direct sunlight.
   (c) 81 mW of DC power in direct sunlight.
   (d) 243 mW of DC power in direct sunlight.
   (e) 729 mW of DC power in direct sunlight.

23. Suppose a single solar cell is capable of supplying exactly 81 mW of DC power when exposed to direct sunlight. If 3 of these cells are connected in parallel, the resulting set of solar cells can theoretically provide

   (a) 9 mW of DC power in direct sunlight.
   (b) 27 mW of DC power in direct sunlight.
   (c) 81 mW of DC power in direct sunlight.
   (d) 243 mW of DC power in direct sunlight.
   (e) 729 mW of DC power in direct sunlight.

24. Which of the following sources of DC power can make use of hydrogen as the chemical from which the power is derived?

   (a) an alkaline cell.
   (b) a zinc-carbon cell.
   (c) a NICAD battery.
   (d) a photovoltaic cell.
   (e) a fuel cell.

25. Electrical current is expressed in terms of

   (a) the rate of change in voltage between two points per unit time.
   (b) the number of charge carriers passing a point per unit time.
   (c) the way in which the resistance varies over time.

(d)   resistance divided by electrical charge.

(e)   the rate of conductance increase per unit time.

26.   In Fig. Test1-3, the five symbols at the top with the arrows represent

   (a)   conventional dry cells.

   (b)   fuel cells.

   (c)   light bulbs.

   (d)   photovoltaic cells.

   (e)   potentiometers.

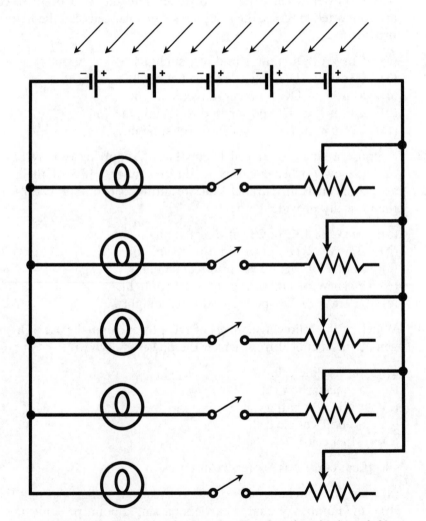

Fig. Test1-3.   Illustration for Part One Test Questions 26 through 33.

27. In Fig. Test1-3, the five symbols at the right with the zig-zaggy lines represent

    (a) conventional dry cells.
    (b) fuel cells.
    (c) light bulbs.
    (d) photovoltaic cells.
    (e) potentiometers.

28. In Fig. Test1-3, the five symbols down the middle with the single arrows and the pairs of little circles represent

    (a) fuel cells.
    (b) resistors.
    (c) switches.
    (d) motors.
    (e) fuses.

29. In Fig. Test1-3, the five symbols consisting of large circles with ovals inside represent

    (a) light bulbs.
    (b) batteries.
    (c) photovoltaic cells.
    (d) fuses.
    (e) fuel cells.

30. Suppose that all five of the components at the top of the circuit in Fig. Test1-3 are turned around, so that the + and − terminals are interchanged. How will this affect the performance of the whole arrangement?

    (a) It won't have any practical effect.
    (b) It will burn out all the components whose symbols consist of large circles with ovals inside.
    (c) It will burn out all the components whose symbols have the zig-zaggy lines.
    (d) It will burn out all the components whose symbols are at the top of the circuit.
    (e) It will make the circuit non-functional, but won't burn anything out.

31.   In Fig. Test1-3, the switches are all

(a)   open.
(b)   connected in parallel with potentiometers.
(c)   connected in parallel with light bulbs.
(d)   closed.
(e)   reversed in polarity.

32.   Suppose one of the components at the top of the circuit in Fig. Test1-3 is removed, leaving an open circuit in its place. How will this affect the performance of the whole arrangement?

(a)   It won't have any practical effect.
(b)   It will burn out all the components whose symbols consist of large circles with ovals inside.
(c)   It will burn out all the components whose symbols have the zig-zaggy lines.
(d)   It will burn out all the components whose symbols are at the top of the circuit.
(e)   It will make the circuit non-functional, but won't burn anything out.

33.   Suppose that the component at the extreme lower right of the circuit in Fig. Test1-3 is removed, leaving an open circuit in its place. How will this affect the performance of the whole arrangement?

(a)   It will make one of the bulbs fail to glow, regardless of the status of the switches.
(b)   It will make all the bulbs glow, regardless of the status of the switches.
(c)   It will make all the bulbs fail to glow, regardless of the status of the switches.
(d)   It will make all the bulbs glow more brightly if any of the switches is closed.
(e)   It will make all the bulbs burn out if any of the switches is closed.

34.   Suppose two electrically charged objects are brought near each other, and an electrostatic force is observed that tends to push them apart. If the objects are brought closer and closer together, the electrostatic force will

(a)   become attractive once the objects come within a certain distance of each other.

(b) decrease in direct proportion to the separation distance, as that distance decreases.

(c) decrease in proportion to the square of the separation distance, as that distance decreases.

(d) remain the same as the separation distance decreases.

(e) None of the above

35. A frequency of 150,000 Hz is the equivalent of

(a) 0.015 MHz.

(b) 0.15 MHz.

(c) 1.5 MHz.

(d) 15 MHz.

(e) None of the above

36. Which of the following equations is true pertaining to the circuit in Fig. Test1-4? The variables I_1, I_2, I_3, and I_4 represent currents, all in amperes.

(a) $I_1 + I_2 = I_3 + I_4$

(b) $I_1 = I_2 + I_3 + I_4$

(c) $I_1 + I_2 + I_3 = I_4$

(d) $I_1 = I_2 = I_3 = I_4$

(e) We can't come up with any equation relating these currents, unless we know the values of the resistors.

37. Suppose that in the circuit of Fig. Test1-4, all the resistors have values of 100 Ω. Suppose $I_1 = 70$ mA and $I_2 = 100$ mA. What is I_4?

(a) 30 mA

(b) 70 mA

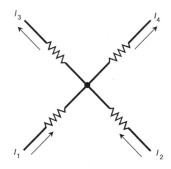

Fig. Test1-4.   Illustration for Test Questions 36 and 37.

(c)   100 mA
(d)   170 mA
(e)   We can't determine I_4 without more information.

38.   Suppose that 6 flashlight cells, each producing 1.5 V, are connected in series in plus-to-plus and minus-to-minus fashion (that is, so that any two terminals that touch each other have the same polarity). What is the net voltage of the combination?

(a)   0 V
(b)   1.5 V
(c)   4.5 V
(d)   9 V
(e)   More information is necessary to answer this question.

39.   Suppose that 5 (instead of 6) flashlight cells, each producing 1.5 V, are connected in series in plus-to-plus and minus-to-minus fashion (that is, so that any two terminals that touch each other have the same polarity). What is the net voltage of the combination?

(a)   0 V
(b)   1.5 V
(c)   4.5 V
(d)   9 V
(e)   More information is necessary to answer this question.

40.   A high EMF can exist even if no current flows. This is called

(a)   alternating voltage.
(b)   static electricity.
(c)   coulomb effect.
(d)   ohmic voltage.
(e)   dynamic charge potential.

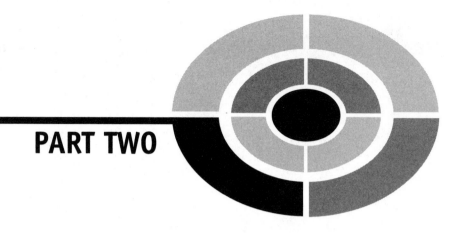

PART TWO

AC Electricity

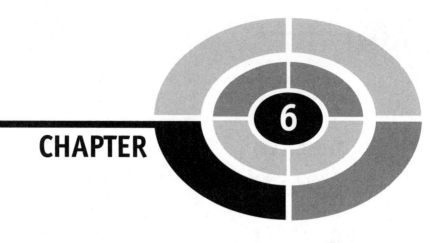

CHAPTER

6

Fundamentals of AC

Alternating current (AC) is electric current in which the charge carriers (usually electrons) reverse direction at regular intervals. In an *AC voltage*, the polarity of the voltage between two points reverses periodically. Appliances that make use of AC are said to consume *AC power*, and over time this accumulates as *AC energy*.

How a Wave Alternates

In an *AC wave*, the alternation rate can be expressed in terms of how long a complete cycle takes, or as the number of complete cycles per unit time.

INSTANTANEOUS AMPLITUDE

The *instantaneous amplitude* of an AC wave is the current, voltage, or power at some exact moment, or instant, in time. The instantaneous amplitude of the AC from a utility outlet constantly and rapidly changes, in contrast to the instantaneous amplitude of the DC from a battery, which does not.

The way in which the instantaneous amplitude varies depends on the shape of the graph of the wave. This *wave shape* is also called the *waveform*.

Instantaneous amplitudes are represented by individual points on wave graphs.

THE CYCLE

When an AC wave is drawn as a graph of current, voltage, or power versus time, a *cycle* is the part of the wave between any point on the graph and the same point on the next alternation. Figure 6-1 shows some points that are commonly used to mark the instant where a single cycle starts and ends. The AC wave in this illustration is a *sine wave*, representing the most perfect possible AC wave. It gets this name because it looks like a graph of the sine function in trigonometry.

CRESTS AND TROUGHS

In Fig. 6-1A, two successive *wave crests* are shown. Crests are points at which the wave is strongest in a specified direction defined as positive. The time required for a single cycle to complete itself corresponds to the distance in the graph between these two points.

Two successive *wave troughs* are also shown in Fig. 6-1A. Troughs are points at which the wave is strongest opposite the direction defined as positive. Again, the time for a cycle is portrayed as the distance between two successive wave troughs.

ZERO POINTS

The time required for one cycle of a wave to complete itself can also be measured between any two successive points at which the wave crosses the *zero axis*. These points correspond to a momentary absence of any current or voltage. Figure 6-1B shows examples of a *positive-going zero point* and a *negative-going zero point*.

PERIOD

The length of time between successive wave crests, troughs, positive-going zero points, or negative-going zero points is called the *period* of the AC wave. Period is denoted by the uppercase italic letter *T*. The standard unit for the period is the second, abbreviated by the lowercase, non-italic letter s.

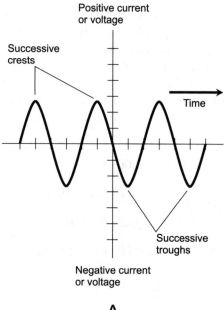

A

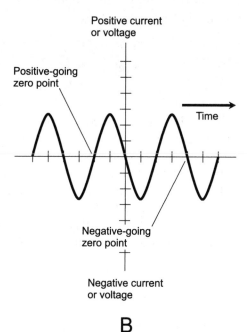

B

Fig. 6-1.   (A) Successive crests and troughs in a wave can be used to define a cycle. (B) Zero points in a wave can be used to define a cycle.

In theory, an AC wave can have a period as short as a tiny fraction of a second, or as long as millions of years! In most electrical and electronics circuits, T is a fraction of a second. The period of a typical household utility AC wave in the United States is 1/60 s. Some radio signals have periods less than 0.000000001 s, or 10^{-9} s.

Period is often expressed in fractions of a second such as *milliseconds* (ms), *microseconds* (µs), *nanoseconds* (ns), and *picoseconds* (ps). The relationships among these time units are:

$$1\,\text{ms} = 0.001\,\text{s}$$

$$1\,\text{µs} = 0.001\,\text{ms} = 10^{-6}\,\text{s}$$

$$1\,\text{ns} = 0.001\,\text{µs} = 10^{-9}\,\text{s}$$

$$1\,\text{ps} = 0.001\,\text{ns} = 10^{-12}\,\text{s}$$

FREQUENCY

The *frequency* of an AC wave, denoted by the lowercase italic letter f, is the reciprocal of the period. That is:

$$f = 1/T$$
$$\text{and}$$
$$T = 1/f$$

In the olden days, frequency was specified in units called *cycles per second*, abbreviated cps. High frequencies were expressed in *kilocycles, megacycles,* or *gigacycles,* representing thousands, millions or billions (thousand-millions) of cycles per second. Now, the standard unit of frequency is known as the *hertz,* abbreviated Hz. The hertz is exactly the same thing as the cycle per second. Thus, 1 Hz = 1 cps, 10 Hz = 10 cps, and 57,338,271 Hz = 57,338,271 cps.

High frequencies are expressed in *kilohertz* (kHz), *megahertz* (MHz) *gigahertz* (GHz), or *terahertz* (THz). The relationships are:

$$1\,\text{kHz} = 1000\,\text{Hz}$$

$$1\,\text{MHz} = 1000\,\text{kHz} = 10^{6}\,\text{Hz}$$

$$1\,\text{GHz} = 1000\,\text{MHz} = 10^{9}\,\text{Hz}$$

$$1\,\text{THz} = 1000\,\text{GHz} = 10^{12}\,\text{Hz}$$

? **PROBLEM 6-1**

You are told that an old computer, on sale at a pawn shop, has a micropro-cessor clock speed of 500 MHz. What is this frequency, f, in hertz? Express the answer as a number written out in full, and also as a number in power-of-10 notation.

✔ **SOLUTION 6-1**

Note that 1 MHz $= 10^6$ Hz, or 1,000,000 Hz. Therefore, 500 MHz is 500 times this:

$$500\,\text{MHz} = (500 \times 1,000,000)\,\text{Hz}$$
$$= 500,000,000\,\text{Hz}$$
$$= 5 \times 10^8\,\text{Hz}$$

? **PROBLEM 6-2**

What is the period T of the wave representing the microprocessor clock signal in the situation of Problem 6-1? Express the answer in seconds, and also in nanoseconds.

✔ **SOLUTION 6-2**

In order to find the period in seconds, we must divide 1 by 500,000,000. A calculator can be useful here if it can display enough digits:

$$T = (1/500,000,000)\,\text{s}$$
$$= 0.000000002\,\text{s}$$
$$= 2 \times 10^{-9}\,\text{s}$$
$$= 2\,\text{ns}$$

The Shape of the Wave

People usually think of a smooth, undulating wave when they think of AC, but there are infinitely many different shapes an AC wave can have. Here are some of the simplest and the most common forms.

SINE WAVE

In a sine wave, the direction of the current reverses at regular intervals, and the current-versus-time curve is shaped like the trigonometric *sine function*. A sine wave, if displayed on an oscilloscope or plotted as a graph, looks like the heavy, wavy curve in Fig. 6-1A or Fig. 6-1B. A wave of this sort is sometimes said to be *sinusoidal*.

Any AC wave that consists of a single, unvarying, defined frequency is sinusoidal. Conversely, any perfect sinusoidal current is entirely concentrated at one single, unvarying, defined frequency.

SQUARE WAVE

When graphed as a function of time, a *square wave* looks like a pair of parallel, dashed lines, one with positive polarity and the other with negative polarity. The transitions between the negative and the positive polarities take place instantaneously, so in theory they should not show up. But often, these transitions are portrayed as vertical lines in illustrations of square waves, so the graph looks like the heavy "curve" in Fig. 6-2.

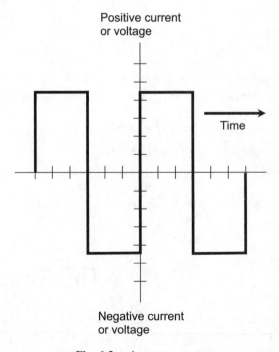

Fig. 6-2.   A square wave.

SAWTOOTH WAVES

Some AC waves rise and/or fall in straight, but diagonal, lines as seen on an oscilloscope screen. The slope of the line indicates how fast the amplitude is changing. Such waves are called *sawtooth waves* because of their appearance. Figure 6-3A shows an example of a sawtooth wave with a finite, measurable *rise* and a practically instantaneous *decay*. Figure 6-3B shows a sawtooth wave with a practically instantaneous rise and a finite, measurable decay.

COMPLEX WAVES

The shape of an AC wave can be complicated, but as long as it has a definite period, and as long as the polarity keeps switching back and forth between positive and negative, it is true AC. With some waves, it can be difficult, or almost impossible, to determine the period, because the wave has two or more *components* that have different periods and, therefore, different frequencies.

[?] **PROBLEM 6-3**

In Fig. 6-3A, suppose each horizontal division represents 1 ns. What is the period of the wave shown?

[✔] **SOLUTION 6-3**

The period, T, is the time between successive crests, troughs, positive-going zero points, or negative-going zero points. There are 3 horizontal divisions between successive crests or between successive troughs. Therefore, $T = 3$ ns.

[?] **PROBLEM 6-4**

What is the frequency of the wave shown in Fig. 6-3A? Express the answer in hertz, and also in megahertz.

[✔] **SOLUTION 6-4**

We know that the period, T, is 3 ns. This is 3×10^{-9} s, or 0.000000003 s. Using a calculator, the approximate frequency, f, is determined by finding

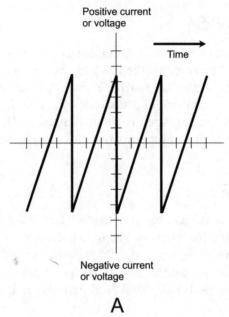

A

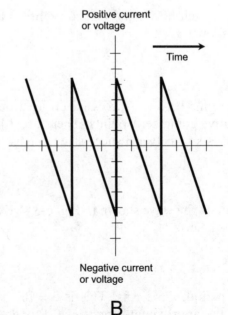

B

Fig. 6-3. (A) An example of a sawtooth wave. This has a slow rise and a fast decay. (B) A sawtooth wave with a fast rise and a slow decay.

the reciprocal of T:

$$f = (1/T)\text{Hz}$$
$$= 1/(0.000000003)\text{Hz}$$
$$= 333,333,333\text{Hz}$$
$$= 3.33 \times 10^8\text{Hz (approximately)}$$

To find the frequency in megahertz, remember that $1\,\text{MHz} = 10^6\,\text{Hz} = 1,000,000\,\text{Hz}$. Therefore, the frequency of the wave shown in Fig. 6-3A, assuming each horizontal division represents 1 ns, is approximately 333 MHz.

The Strength of the Wave

Depending on the quantity being measured, the amplitude of an AC wave can be expressed as amperes (for current), volts (for voltage), or watts (for power). There are several ways in which AC amperes, volts, or watts can be quantified.

PEAK AMPLITUDE

The *peak amplitude* of an AC wave is the maximum extent, either positive or negative, that the instantaneous amplitude attains. In many waves, the positive and negative peak amplitudes are the same. But sometimes they differ. Peak is often abbreviated pk. You might see the amplitude of an AC wave expressed as 4.5 A pk, for example.

Figure 6-4A illustrates the concept of peak current or voltage for a pure AC sine wave. The positive or negative peak amplitudes of an AC wave do not depend on the period or the frequency.

PEAK-TO-PEAK AMPLITUDE

The *peak-to-peak amplitude* of a wave is the difference between the positive peak amplitude and the negative peak amplitude, taking polarity into account. Peak-to-peak (pk–pk) is a way of expressing how much the wave level "swings" during the cycle. If the positive and negative peak amplitudes are the same, then the peak-to-peak amplitude is twice the peak amplitude.

Figure 6-4B shows the notion of peak-to-peak current or voltage for a pure AC sine wave. Peak current has a direction assigned to it, and peak voltage has a polarity. But peak-to-peak quantities do not have direction or polarity. Thus, after calculating the peak-to-peak current or voltage, you should remove any directional expression or polarity sign. The peak-to-peak amplitude of an AC wave does not depend on the period or the frequency.

AVERAGE AMPLITUDE

The *average amplitude* of an AC wave is the instantaneous amplitude, mathematically averaged over exactly one complete cycle, or a whole number of complete cycles. In a pure AC sine wave, the average amplitude is zero because the wave is positive half the time and negative half the time, and the "positivity" and "negativity" exactly cancel each other out. The same is true for a square AC wave, or for any symmetrical AC wave, as long as there is no DC superimposed on it.

An irregular AC wave sometimes has an average amplitude of zero, but often this is not the case. The average amplitude for an irregular wave depends on the shape of the wave, and also on the positive and negative peak amplitudes. It does not depend on the period or the frequency. There are infinitely many possible irregular waveforms, and some of them are incredibly complex. We won't get into the theory of irregular waveforms in this course.

If an AC sine wave has a *DC component* along with it, then the average amplitude of the wave is the same as the DC component. For example, if a common utility AC voltage source is connected in series with a 45 V DC battery, the resulting output has an average voltage of +45 V or −45 V, depending on which way the battery is connected. Situations of this kind are rare in ordinary electrical circuits, but they occasionally arise in electronic devices.

ROOT-MEAN-SQUARE AMPLITUDE

Often, it is necessary to express the *effective amplitude* or the *DC-equivalent amplitude* of an AC wave. This is almost never the same thing as average amplitude! The effective amplitude of an AC wave is the voltage, current or power that a DC source would have to produce in order to have the same practical effect. When people say that a wall outlet supplies 117 V, they mean 117 effective volts.

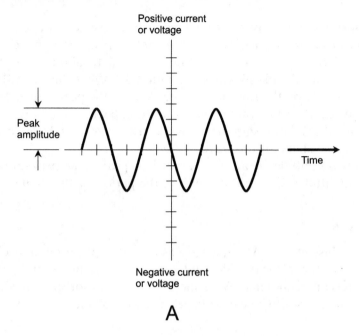

A

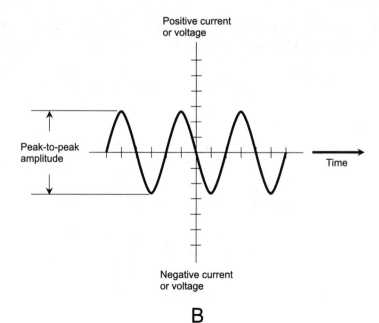

B

Fig. 6-4.   (A) Peak (pk) amplitude for a sine wave. (B) Peak-to-peak (pk–pk) amplitude for a sine wave.

The most common way of expressing effective AC amplitude is called the *root-mean-square amplitude*. This is abbreviated rms. For a perfect sine wave with no DC component, the rms value is equal to about 0.354 times the pk–pk value, and the pk–pk value is about 2.828 times the rms value. For a perfect square wave, the rms value is the same as the peak value. The pk–pk value is twice the rms value and twice the peak value. For other AC waveforms, the relationship between the rms value and the peak value depends on the shape of the wave.

The rms amplitude of an AC wave, like peak, and peak-to-peak, and average amplitudes, does not depend on the period or the frequency.

? PROBLEM 6-5

Figure 6-5 illustrates an AC sine wave as it might appear on an oscilloscope. Voltage is on the vertical scale, and time is on the horizontal scale. Each vertical division represents 5 V. Time divisions are not quantified. What is the approximate positive peak voltage? The approximate negative peak voltage?

✔ SOLUTION 6-5

A close examination shows that the peak positive voltage is about +23 V, and the peak negative voltage is about −23 V.

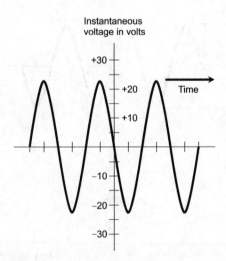

Fig. 6-5.   Illustration for Problems 6-5 through 6-8.

|?| **PROBLEM 6-6**

What is the approximate peak-to-peak voltage of the AC wave shown in Fig. 6-5?

|✔| **SOLUTION 6-6**

The peak-to-peak voltage, E_{pk-pk}, is the difference between the positive and the negative peak voltages. This can be found by subtracting $-23\,V$ from $+23\,V$ and then removing the sign (because peak-to-peak voltages don't have polarities):

$$E_{pk-pk} = +23 - (-23)V$$
$$= +23 + 23V$$
$$= 46V$$

|?| **PROBLEM 6-7**

What is the average voltage of the AC wave shown in Fig. 6-5?

|✔| **SOLUTION 6-7**

The average voltage is $0\,V$, assuming the positive and negative peak voltages are equal and opposite, and assuming (as we are told in Problem 6-5) that the wave is a sine wave.

|?| **PROBLEM 6-8**

What is the rms voltage of the AC wave shown in Fig. 6-5?

|✔| **SOLUTION 6-8**

For a sine wave, the rms value is approximately 0.354 times the peak-to-peak value, as long as there is no DC component. In this example, there is no DC component, so the rms voltage, E_{rms}, is approximately:

$$E_{rms} = (0.354 \times 46)V$$
$$= 16.3V$$

How AC Is Produced

There are several ways of obtaining AC electricity. Electric generators are the most common means of producing utility AC. Another scheme involves the conversion of DC to AC using specialized switches and transformers.

THE AC GENERATOR

An *AC generator* is an electromechanical device designed to produce electricity using a coil and magnetic field. Whenever an electrical conductor, such as a wire, moves in a magnetic field, electric current is induced in the conductor. A generator can consist of a rotating magnet inside a coil of wire, or a rotating coil inside a magnet or pair of magnets. The rotating shaft can be driven by any sort of mechanical force, such as falling water, flowing rivers, prevailing winds, ocean currents, nuclear reactions, heat from inside the earth, steam under pressure, or explosive combustion of fossil fuels. The output of such a generator is always AC.

Small portable gasoline-powered generators, capable of producing a few kilowatts, can be purchased in department stores. Larger generators can produce enough electricity to supply a house or building. The most massive electric generators, found in power plants, are as large as a house, can produce several megawatts, and can provide sufficient electricity for a community.

When AC is generated by a rotating magnet in a coil of wire, or by a rotating coil of wire inside a powerful magnet, AC voltage appears between the ends of the length of wire. The voltage depends on the strength of the magnet, the number of turns in the wire coil, and the speed at which the magnet or coil rotates. The AC frequency depends only on the speed of rotation. Normally, for utility AC, this speed is 3600 revolutions per minute (rpm), or 60 revolutions per second (rps). This produces an output frequency of 60 Hz. In some countries the rotation speed is 2500 rpm or 50 rps, producing an AC output frequency of 50 Hz.

GENERATOR EFFICIENCY

It doesn't take very much rotational force, or *torque*, to turn the shaft of a generator when there is nothing connected to its output. But when a *load*—something that draws current, such as a light bulb or an electric heater—is connected to an AC generator, it becomes more difficult to turn the shaft.

The more electrical power that is demanded from a generator, the greater will be the amount of mechanical power required to drive it. This is why you cannot connect a generator to a stationary bicycle and pedal a city into electrification. The electrical power that comes out of a generator can never be more than the mechanical power driving it. In fact, there is always some energy lost, mainly as heat in the generator hardware. If you connect a generator to a stationary bicycle, your legs might provide enough power to run a small radio, but nowhere near enough to supply a household.

Generator efficiency is the ratio of the power output to the mechanical power input, both measured in the same units (such as watts or kilowatts). The ratio can be multiplied by 100 to get a percentage. If P_D is the mechanical power that drives a generator and P_{AC} is the AC output power in the same units, then the efficiency (*Eff*), as a ratio, can be calculated using this formula:

$$Eff = P_{AC}/P_D$$

The efficiency (*Eff*%), as a percentage, is

$$Eff_\% = 100(P_{AC}/P_D)\%$$

THE POWER INVERTER

A *power inverter*, sometimes called a *chopper power supply*, is a circuit that delivers high-voltage AC from a low-voltage DC source. The input is typically 12 to 14 V DC, and the output is usually 110 to 130 V AC. Small power inverters can be obtained in department stores. These can operate small devices such as lamps and low-power hi-fi radios, but they aren't designed to power large appliances, households, or businesses.

Figure 6-6 is a simplified block diagram of a power inverter. The chopper opens and closes a *switching transistor*. This interrupts the battery current, producing pulsating DC. The transformer converts the pulsating DC to AC, and also steps up the voltage. If the battery is rechargeable, solar panels can be used to replenish its charge and provide a long-term source of utility power.

The output of a low-cost power inverter is generally not a good sine wave, but more closely resembles a square wave. Square-wave choppers are inexpensive and easy to manufacture. For this reason, low-cost power inverters do not always work well with appliances that need a nearly perfect, 60-Hz sine-wave source of AC power. Sophisticated (and expensive) inverters produce fairly good sine waves, and have a frequency close to 60 Hz. They

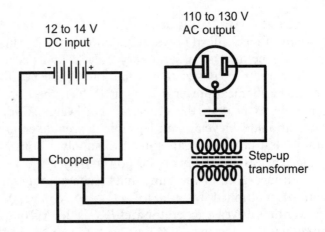

Fig. 6-6. A power inverter for converting low-voltage DC to utility AC.

are a good investment if they are to be used with sensitive equipment such as computers.

EDISON VERSUS AC

Thomas Edison is said to have favored DC over AC for electrical power transmission in the early 1900s, when electric utilities were first being planned and developed. His colleagues argued that AC would work better, and they prevailed. But there is at least one advantage to DC in electric power transmission. This becomes apparent in long-distance power lines. Direct currents, at extremely high voltages, are transported more efficiently than alternating currents. The wire has less effective resistance with DC than with AC, and there is less energy lost in the magnetic fields that always surround current-carrying wires.

? **PROBLEM 6-9**

Mechanical power is often measured in terms of units called *horsepower*, symbolized hp. A mechanical power level of 1 hp is equivalent to approximately 746 W. Suppose it takes 2 hp of mechanical energy to turn a generator that puts out 1000 W of AC electricity. What is the efficiency of this generator as a ratio? As a percentage?

✔ **SOLUTION 6-9**

In this case, $P_D = (746 \times 2)$ W $= 1492$ W. We are told that $P_{AC} = 1000$ W. Therefore, the efficiency *Eff*, as a ratio, is:

$$Eff = P_{AC}/P_D$$
$$= 1000/1492$$
$$= 0.67$$

The efficiency $Eff_\%$, as a percentage, is:

$$Eff_\% = 100(P_{AC}/P_D)\%$$
$$= 100(1000/1492)\%$$
$$= (100 \times 0.67)\%$$
$$= 67\%$$

Quiz

This is an "open book" quiz. You may refer to the text in this chapter. A good score is 8 correct answers. Answers are in Appendix 1.

1. Suppose you see an AC sine wave on an oscilloscope, and the oscilloscope screen looks like Fig. 6-7. You have the scope set so each horizontal division represents exactly 5 ms and each vertical division represents exactly 1 V. The period of this wave is approximately

    (a)   15 ms
    (b)   30 ms
    (c)   3 ms
    (d)   6 ms

2. If each horizontal division in Fig. 6-7 represents exactly 5 ms and each vertical division represents exactly 1 V, then the frequency of the wave is approximately

    (a)   33 Hz
    (b)   67 Hz
    (c)   167 Hz
    (d)   333 Hz

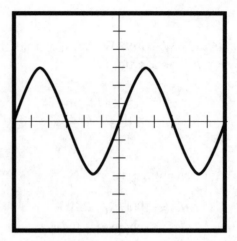

Fig. 6-7. Illustration for Quiz Questions 1 through 4.

3. If each horizontal division in Fig. 6-7 represents exactly 5 ms and each vertical division represents exactly 1 V, then the positive peak voltage of the wave, assuming there is no DC component superimposed, is approximately

    (a) +3 V
    (b) +6 V
    (c) +15 V
    (d) +30 V

4. If each horizontal division in Fig. 6-7 represents exactly 5 ms and each vertical division represents exactly 1 V, then the rms voltage of the wave, assuming it is a perfect sine wave and there is no DC component superimposed, is approximately

    (a) 1.06 V rms
    (b) 2.12 V rms
    (c) 5.3 V rms
    (d) 10.6 V rms

5. If the period of an AC sine wave doubles, and if nothing else about it changes, then the frequency

    (a) doubles.
    (b) is cut in half.
    (c) quadruples.
    (d) stays the same.

6. If the period of a perfect square AC wave doubles, and if nothing else
    about it changes, then the ratio of its rms amplitude to its peak-to-
    peak amplitude

    (a)   doubles.
    (b)   is cut in half.
    (c)   quadruples.
    (d)   stays the same.

7. A generator requires 800 kW of mechanical power to operate, and is
    capable of producing 500,000 W of AC power. What is the efficiency
    of this generator?

    (a)   625%
    (b)   160%
    (c)   62.5%
    (d)   16%

8. The average amplitude of an irregular AC wave

    (a)   is always zero.
    (b)   depends on the frequency.
    (c)   depends on the period.
    (d)   depends on the waveform.

9. If an AC sine wave has a period of approximately 10 µs, then an AC
    square wave of the same frequency has a period of approximately

    (a)   10 µs.
    (b)   20 µs.
    (c)   3.54 µs.
    (d)   2.8 µs.

10. The instantaneous amplitude of an AC wave

    (a)   has a constant frequency.
    (b)   consists of only one frequency.
    (c)   has a constant period.
    (d)   constantly varies.

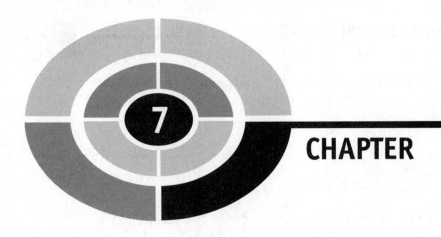

CHAPTER

Electricity in the Home

In this chapter, we'll look at some of the most important characteristics of standard utility AC. This is the "electricity" used in homes, and with which everyone is more or less familiar.

Phase

Phase is an expression of a point in time during an AC cycle. Phase can also express the relative time displacement between two AC waves. When comparing two waves, the *phase difference* can be defined if, but only if, the waves have precisely the same frequency.

DEGREES OF PHASE

One method of defining the phase of an AC cycle is to divide it into 360 equal parts called *degrees* or *degrees of phase*. The value 0° is assigned to the point in the cycle where the magnitude is zero and positive-going. The same point on the next cycle is given the value 360°. Halfway through the cycle is 180°, a quarter of the way through the cycle is 90°, and three-quarters of the way through the cycle is 270°. Figure 7-1 illustrates this concept for a sine wave.

COINCIDENCE AND OPPOSITION

The term *phase coincidence* means that two waves are in lock-step with each other. This situation is shown in Fig. 7-2A for two waves having different amplitudes. The phase difference in this case is 0°.

If two sine waves are in phase coincidence, the peak-to-peak amplitude of the resultant wave, which is also a sine wave, is equal to the sum of the peak-to-peak amplitudes of the two constituent waves. The phase of the resultant is the same as that of the constituent waves.

When two AC sine waves begin exactly 1/2 cycle, or 180°, apart, they are said to be in *phase opposition* (Fig. 7-2B). In this situation, people sometimes say that the waves are *180° out of phase*.

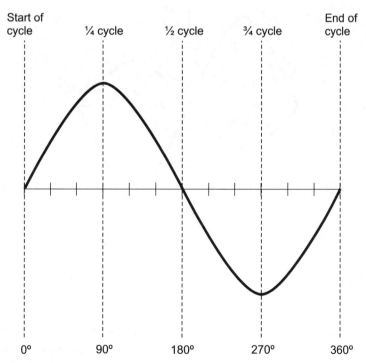

Fig. 7-1. Degrees of phase in a sine wave are expressions of the portion of the cycle that has passed since its starting time.

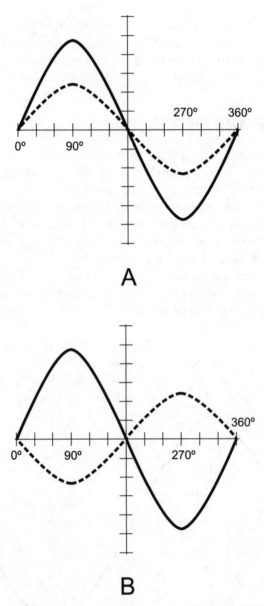

Fig. 7-2. At A, graphic illustration of sine waves in phase coincidence. At B, graphic illustration of sine waves in phase opposition.

If two sine waves have the same amplitude and are in phase opposition, they cancel each other out. The instantaneous amplitudes are equal and opposite at every point in time. If two sine waves have different amplitudes and are in phase opposition, the peak-to-peak value of the resultant wave, which is also a sine wave, is equal to the difference between the peak-to-peak values of the two constituent waves, and the phase of the resultant is the same as the phase of the stronger of the two constituents.

SINGLE-PHASE AC

Single-phase AC consists of a lone sine wave, clean and uncomplicated, as in Fig. 7-1. This is the sort of AC available at standard wall outlets intended for small appliances such as lamps, TV sets, and computers. In most parts of the United Stated, the rms voltage is standardized at 117 V, but it can vary somewhat above and below this level depending on power demand at the time, your location, and to some extent at the whims of your local electric utility. The positive and negative peak voltages are 1.414 times the rms voltage, or about +165 V pk+ and −165 V pk−. The peak-to-peak voltage is about twice the positive or negative peak voltage, or 330 V pk–pk.

3-PHASE AC

Electricity is transmitted over major power lines as three sine waves, each having the same peak voltage. Each wave differs in phase by 120°, or 1/3 of a cycle, with respect to the other two. This is called a *3-phase AC*. Figure 7-3 shows what a 3-phase AC electrical waveform looks like when plotted as a graph of instantaneous voltage versus time.

The horizontal axis in this figure is graduated in degrees for phase 1 (solid curve). Phase 2 (coarse dashed curve) comes 1/3 of a cycle later than Phase 1, and phase 3 comes 2/3 of a cycle later than phase 1.

There are advantages to 3-phase AC over single-phase AC. Each wave provides its own energy, independent of the other two. When a 3-phase AC wave enters a household or business, up to 3 different, and independent, single-phase electrical circuits can be built. This helps to distribute the power demand around the various rooms and appliances. In addition, a 3-phase AC is more efficiently transmitted along power lines than a single-phase AC.

? **PROBLEM 7-1**

Suppose two AC waves are exactly in phase. Call them wave X and wave Y. There is no DC component on either wave. The peak-to-peak voltage of wave X is 100 V, and the peak-to-peak voltage of wave Y is 150 V. What is the peak-to-peak voltage of the composite wave, $Z = X + Y$? The positive peak voltage? The negative peak voltage? The rms voltage?

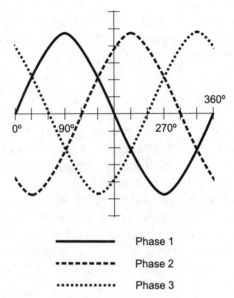

Fig. 7-3. Graphic illustration of 3-phase AC.

✔ **SOLUTION 7-1**

The peak-to-peak voltage of wave Z is equal to the sum of the peak-to-peak voltages of waves X and Y:

$$E_{Zpk-pk} = E_{Xpk-pk} + E_{Ypk-pk}$$
$$= 100Vpk-pk + 150Vpk-pk$$
$$= 250Vpk-pk$$

The positive peak voltage of wave Z is equal to half the peak-to-peak voltage of wave Z. Therefore:

$$E_{Zpk+} = (E_{Zpk-pk})/2$$
$$= +(250/2)Vpk+$$
$$= +125Vpk+$$

The negative peak voltage of wave Z is equal to half the peak-to-peak voltage of wave Z, multiplied by −1. Therefore:

$$E_{Zpk-} = -(E_{Zpk-pk})/2$$
$$= -(250/2) Vpk-$$
$$= -125 Vpk-$$

The rms voltage of wave Z is equal to approximately 0.354 times the peak-to-peak voltage of wave Z. Therefore:

$$E_{Zrms} = 0.354E_{Zpk-pk}$$
$$= (0.354 \times 250)\text{Vrms}$$
$$= 88.5 \text{ Vrms}$$

? **PROBLEM 7-2**

Suppose two AC waves are exactly out of phase. Call them, again, wave X and wave Y. There is no DC component on either wave. The peak-to-peak voltage of wave X is 100 V, and the peak-to-peak voltage of wave Y is 150 V. What is the peak-to-peak voltage of the composite wave, Z = X + Y? The positive peak voltage? The negative peak voltage? The rms voltage?

✔ **SOLUTION 7-2**

The peak-to-peak voltage of wave Z is equal to the *absolute value* of the difference between the peak-to-peak voltages of waves X and Y. The absolute value of a number is equal to that number if the number is positive, and equal to −1 times that number if the number is negative. Absolute value is symbolized by placing vertical lines on either side of a quantity. Therefore:

$$E_{Zpk-pk} = \left| E_{Xpk-pk} - E_{Ypk-pk} \right|$$
$$= \left| 100 \text{ V pk} - \text{pk} - 150 \text{ V pk} - \text{pk} \right|$$
$$= 50 \text{ V pk} - \text{pk}$$

The positive peak voltage of wave Z is equal to half the peak-to-peak voltage of wave Z. Therefore:

$$E_{Zpk+} = (E_{Zpk-pk})/2$$
$$= +(50/2) \text{ V pk}+$$
$$= +25 \text{ V pk}+$$

The negative peak voltage of wave Z is equal to half the peak-to-peak voltage of wave Z, times −1. Therefore:

$$E_{Zpk-} = -(E_{Zpk-pk})/2$$
$$= -(50/2) \text{ V pk}-$$
$$= -25 \text{ V pk}-$$

The rms voltage of wave Z is equal to approximately 0.354 times the peak-to-peak voltage of wave Z. Therefore:

$$E_{Zrms} = 0.354(E_{Zpk-pk})$$
$$= (0.354 \times 50) \text{ V rms}$$
$$= 17.7 \text{ V rms}$$

Transformers

A *transformer* is a device used to change AC voltage. Transformers work by means of a phenomenon called *inductive coupling*. In electrical applications, transformers usually consist of wires wound around cores made of *laminated iron* (thin slabs of iron glued together).

WINDINGS

The *primary* of a transformer is the winding to which electricity is applied. The *secondary* is the winding from which electricity is taken. In a *step-down transformer*, the primary has more turns than the secondary. In a *step-up transformer*, the primary has fewer turns than the secondary. Figure 7-4A shows the schematic symbol for a step-down transformer, and Fig. 7-4B shows the schematic symbol for a step-up transformer. The solid, parallel, vertical lines indicate that the transformers have laminated iron cores.

The size, or *gauge*, of the wire used for a primary winding depends on the amount of current it must carry. The higher the current demand, the larger must the wire be. For a given amount of power, a step-up transformer must use larger primary-winding wire than a step-down transformer. A transformer that steps up the voltage by a large factor must have a primary that consists of very heavy wire.

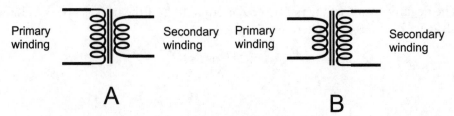

Fig. 7-4. At A, schematic symbol for a step-down AC transformer. At B, schematic symbol for a step-up AC transformer.

The size of the wire in the secondary winding of a transformer depends on the power demand and also on the *turns ratio*, which is defined as the number of turns in the primary divided by the number of turns in the secondary. A step-up transformer (turns ratio less than 1:1) can have a smaller secondary-winding wire than a step-down transformer (turns ratio larger than 1:1) for a given amount of power. The secondary winding of a transformer can have intermediate connections, called *taps*, for obtaining various AC output voltages.

OPERATION

Small transformers are used in power supplies for electronic devices, such as computers and radios. We'll look at power supplies in the next chapter. Medium-sized transformers are used in high-current and/or high-voltage power supplies. Still larger ones provide the utility power to which we are accustomed in our homes and businesses. The largest transformers handle thousands of volts and thousands of amperes, and are used at power-transmission stations. Some of these are as big, and as heavy as a fully loaded truck.

If a transformer has a primary winding with N_{pri} turns and a secondary winding with N_{sec} turns, then the primary-to-secondary turns ratio, T, is defined as follows:

$$T = N_{\text{pri}}/N_{\text{sec}}$$

This ratio determines the voltage ratio between the windings.

For a transformer with a primary-to-secondary turns ratio T and an efficiency of 100% (that is, no power loss), the voltage E_{sec} across the entire secondary winding is related to the voltage E_{pri} across the entire primary winding according to the following equations:

$$E_{\text{pri}} = E_{\text{sec}} \times T$$
$$E_{\text{sec}} = E_{\text{pri}}/T$$

EFFICIENCY

In a real-life transformer, some power is always lost in the coil windings. Some power is also lost in the core. *Conductor losses* occur because of the *ohmic resistance* of the wire that makes up the windings. *Core losses* occur because of *eddy currents* (circulating currents in the iron) and *hysteresis* ("sluggishness" of the iron's response to AC magnetic fields). The efficiency of a transformer is therefore always less than 100%.

Let E_{pri} and I_{pri} represent the primary-winding voltage and current in a hypothetical transformer, and let E_{sec} and I_{sec} represent the secondary-winding voltage and current. In a perfect transformer, the product $E_{pri}I_{pri}$ would be equal to the product $E_{sec}I_{sec}$. But in a real-world transformer, $E_{pri}I_{pri}$ is always greater than $E_{sec}I_{sec}$. The efficiency Eff of a transformer, expressed as a ratio, is:

$$Eff = E_{sec}I_{sec}/(E_{pri}I_{pri})$$

Expressed as a percentage, the efficiency $Eff_\%$ is:

$$Eff_\% = 100E_{sec}I_{sec}/(E_{pri}I_{pri})\%$$

The power P_{loss} dissipated in the transformer windings and core, and thereby wasted as heat, is equal to:

$$P_{loss} = E_{pri}I_{pri} - E_{sec}I_{sec}$$

These formulas are valid for electrical units of volts (V), amperes (A), and watts (W). The primary and secondary voltage and current must also be of the same type (positive/negative peak, peak-to-peak, or rms). If positive or negative peak values are specified, there must be no DC component in either the input or the output. The reason for this is the fact that a transformer is a *DC-blocking* device.

The blocking of DC is an important characteristic of transformers that you should remember. Alternating current can pass through a transformer, but direct current cannot. Therefore, if the input wave to a transformer has a DC component, that component does not appear in the output wave. This property of transformers can be useful in the design of certain electrical devices and systems.

The efficiency of a transformer varies depending on the load connected to the secondary winding. If the current drain is excessive, the efficiency is reduced. Efficiency is also reduced if there is a DC component in the input. Transformers are rated according to the maximum amount of power they can deliver without any serious degradation in efficiency.

? **PROBLEM 7-3**

Suppose the AC input voltage to a step-down transformer is 120 V rms. There is no DC component. This voltage, $E_{pri\text{-}rms}$, appears across the entire primary winding. The primary-to-secondary turns ratio, T, is 5:1. The transformer has a secondary with a tap in the center (Fig. 7-5).

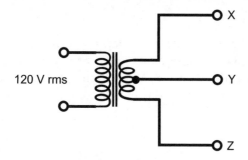

Fig. 7-5. Illustration for Problems 7-3 and 7-4.

What is the rms output voltage, $E_{\text{sec-rms}}$, that appears across the entire secondary winding, that is, between terminals X and Z? Assume the input is a sine wave.

✔ **SOLUTION 7-3**

The rms secondary voltage is 1/5, or 0.2, times the rms primary voltage. This is apparent if we plug numbers into the formula:

$$E_{\text{sec-rms}} = E_{\text{pri-rms}}/T$$
$$= (120/5) \text{ V rms}$$
$$= 24 \text{ V rms}$$

? **PROBLEM 7-4**

What is the rms voltage that appears between terminals X and Y of the secondary in Fig. 7-5, assuming the same turns ratio as in the previous problem? What is the rms voltage that appears between terminals Y and Z? Assume that the input is a sine wave with no DC component.

✔ **SOLUTION 7-4**

The tap, at terminal Y, is in the center of the transformer. This splits the voltage into two equal parts. The voltage between terminals X and Y is therefore (24/2) V rms = 12 V rms, and the voltage between terminals Y and Z is also 12 V rms.

? **PROBLEM 7-5**

Suppose the AC input voltage to a step-up transformer is 12 V rms. This voltage, $E_{pri\text{-}rms}$, appears across the entire primary winding. The primary-to-secondary turns ratio, T, is 1:8. What is the peak-to-peak output voltage, $E_{sec\text{-}pk\text{-}pk}$, that appears across the entire secondary winding? Assume the input is a sine wave.

✔ **SOLUTION 7-5**

The rms (not peak-to-peak) secondary voltage, $E_{sec\text{-}rms}$, is 8 times the rms primary voltage, or 96 V rms. This is apparent if we plug numbers into the formula:

$$E_{sec\text{-}rms} = E_{pri\text{-}rms}/T$$

$$= (12/0.125) \text{ V rms}$$

$$= (12 \times 8) \text{ V rms}$$

$$= 96 \text{ V rms}$$

The peak-to-peak voltage across the secondary, $E_{sec\text{-}pk\text{-}pk}$, is approximately equal to the rms voltage times 2.828:

$$E_{sec\text{-}pk\text{-}pk} = 2.828 \; E_{sec\text{-}rms}$$

$$= (96 \times 2.828) \text{ V pk}-\text{pk}$$

$$= 271 \text{ V pk}-\text{pk}$$

? **PROBLEM 7-6**

Suppose a load is connected to a transformer. The voltage across the primary winding is 120 V rms, and the current through the primary winding is 2.57 A rms. The voltage across the secondary winding is 12.3 V rms, and the current drawn by the load connected across the secondary winding is 19.9 A rms. What is the efficiency of this transformer, expressed as a ratio? As a percentage?

✔ **SOLUTION 7-6**

To determine the efficiency, *Eff*, plug in the numbers to the formula:

$$Eff = E_{sec}I_{sec}/(E_{pri}I_{pri})$$
$$= (12.3 \times 19.9)/(120 \times 2.57)$$
$$= 244.77/308.4$$
$$= 0.794$$

To express the efficiency as a percentage, $Eff_\%$, multiply *Eff* by 100. This gives us the figure $Eff_\% = 79.4\%$.

Ohm's Law, Power, and Energy

Household AC behaves in the same way as DC when it comes to Ohm's Law, power, and energy, as long as the rms AC values are used, and as long as the current and the voltage are exactly in phase with each other. In most household and business appliances, the current is in phase with the voltage.

REACTANCE AND IMPEDANCE

When the current and the voltage are not in phase with each other in an AC circuit, quantities known as *reactance* and *complex impedance* replace ordinary resistance. These phenomena are beyond the scope of this elementary course. If you are interested in them, an excellent reference is *Teach Yourself Electricity and Electronics* (McGraw-Hill, New York). A lighter treatment is given in the "sister" to this book, *Electronics Demystified* (McGraw-Hill, New York).

OHM'S LAW FOR AC

Imagine a circuit containing a source of AC electricity, a load (shown as a resistor), and an AC ammeter that measures the rms current through the resistor (Fig. 7-6). Let E_{rms} stand for the voltage of the AC source (in volts rms), let I_{rms} stand for the current through the load (in amperes rms), and let R stand for resistance of the load (in ohms). Three formulas denote

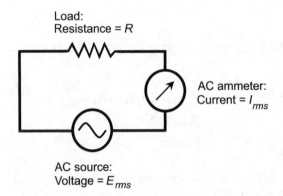

Fig. 7-6.   A basic AC circuit for demonstrating Ohm's Law, power, and energy calculations. Illustration for Problems 7-7 through 7-15.

Ohm's law for AC:

$$E_{rms} = I_{rms}R$$
$$I_{rms} = E_{rms}/R$$
$$R = E_{rms}/I_{rms}$$

You must always use units of volts rms (V rms), amperes rms (A rms), and ohms (Ω) for these formulas to work. If quantities are given in units other than volts rms, amperes rms, and ohms, you must convert to these units and then calculate. After that, you can convert the units back to whatever you like.

? **PROBLEM 7-7**

Suppose the AC source in the circuit of Fig. 7-6 produces 40 V rms, and the resistor has a value of 20 Ω. What is the current?

✔ **SOLUTION 7-7**

Plug in the numbers to the Ohm's Law formula for current, as follows:

$$I_{rms} = E_{rms}/R$$
$$= (40/20) \text{ A rms}$$
$$= 2 \text{ A rms}$$

☐? **PROBLEM 7-8**

Suppose the load in the circuit of Fig. 7-6 has a resistance of $100\,\Omega$, and the measured current is $100\,\text{mA}$ rms. What is the rms AC voltage of the AC source?

☑ **SOLUTION 7-8**

Use the formula $E_{rms} = I_{rms}R$. First, convert the current to amperes: $100\,\text{mA}$ rms $= 100 \times 0.001\,\text{A}$ rms $= 0.1\,\text{A}$ rms. Then plug in the numbers:

$$E_{rms} = I_{rms}R$$
$$= (0.1 \times 100)\text{V rms}$$
$$= 10\,\text{V rms}$$

☐? **PROBLEM 7-9**

If the AC source voltage in the circuit of Fig. 7-6 is $117\,\text{V}$ rms and the ammeter shows $3\,\text{A}$ rms, what is the value of the resistor?

☑ **SOLUTION 7-9**

Use the formula $R = E_{rms}/I_{rms}$, and plug in the values directly, because they are expressed in volts rms and amperes rms:

$$R = E_{rms}/I_{rms}$$
$$= (117/3)\,\Omega$$
$$= 39\,\Omega$$

AC POWER CALCULATIONS

Three formulas can be used to determine the AC power, P (in watts) consumed by the load in the generic circuit of Fig. 7-6:

$$P = E_{rms}I_{rms}$$
$$P = (E_{rms})^2/R$$
$$P = (I_{rms})^2R$$

You must always use units of volts rms (V rms), amperes rms (A rms), and ohms (Ω) for power calculations using these formulas to come out in watts. If any of the quantities is specified in some unit other than these, you must convert before you calculate.

[?] **PROBLEM 7-10**

Suppose the AC source voltage in the circuit of Fig. 7-6 is 60 V rms, and the ammeter reads 5 A rms. What is the power consumed by the load?

[✔] **SOLUTION 7-10**

Plug in the numbers as follows:

$$P = E_{rms}I_{rms}$$
$$= (60 \times 5) \text{ W}$$
$$= 300 \text{ W}$$

[?] **PROBLEM 7-11**

Suppose the AC source voltage in the circuit of Fig. 7-6 is 10 V rms, and the resistance of the load is 50 Ω. What is the power consumed by the load?

[✔] **SOLUTION 7-11**

Plug the values into the formula for power in terms of rms voltage and resistance:

$$P = (E_{rms})^2 / R$$
$$= (10 \times 10/50) \text{ W}$$
$$= (100/50) \text{ W}$$
$$= 2 \text{ W}$$

[?] **PROBLEM 7-12**

Suppose the load in the circuit of Fig. 7-6 has an internal resistance of 10 Ω, and the measured current is through it is 6 A rms. What is the power?

✔ **SOLUTION 7-12**

Use the formula for power in terms of rms current and resistance. Plug in the numbers:

$$E = (I_{rms})^2 R$$
$$= (6 \times 6 \times 10) \text{ W}$$
$$= (36 \times 10) \text{ W}$$
$$= 360 \text{ W}$$

AC ENERGY CALCULATIONS

Three formulas can be used to determine the energy, W (in watt-hours) consumed by the load in the circuit of Fig. 7-6:

$$W = E_{rms} I_{rms} t$$
$$W = (E_{rms})^2 t / R$$
$$W = (I_{rms})^2 R t$$

You must always use units of volts rms (V rms), amperes rms (A rms), ohms (Ω), and hours (h) for energy calculations to come out in watt-hours. If quantities are given in units other than these units, you must convert them before you plug any numbers into the above formulas.

? **PROBLEM 7-13**

Suppose the circuit of Fig. 7-6 is activated for 3 h and then switched off. Suppose the AC voltage source supplies 12 V rms and the ammeter shows 4 A. What is the energy consumed, in watt-hours?

✔ **SOLUTION 7-13**

Use the first of the three formulas above. We're given the quantities in standard units, so there is no need to convert any of them. We can plug in the numbers directly. Here, $E_{rms} = 12$, $I_{rms} = 4$, and $t = 3$. Therefore:

$$W = E_{rms} I_{rms} t$$
$$= (12 \times 4 \times 3) \text{Wh}$$
$$= 144 \text{ Wh}$$

PROBLEM 7-14

Now imagine that we don't know the ammeter reading in the circuit of Fig. 7-6, but we know that the circuit is activated for 15 min and the AC source voltage is 240 V rms. The resistance of the load is known to be 60 Ω. How much energy is consumed by the load during this period of time? Give the answer in watt-hours, and then in kilowatt-hours.

☑ **SOLUTION 7-14**

First, convert the time to hours: 15 min = 0.25 h. Therefore, $t = 0.25$. Then, note that $E_{rms} = 240$ and $R = 60$. Use the second of the above formulas for energy:

$$W = (E_{rms})^2 t/R$$

$$= (240 \times 240 \times 0.25/60) \text{ Wh}$$

$$= 240 \text{ Wh}$$

$$= 0.24 \text{ kWh}$$

?
 PROBLEM 7-15

Suppose we don't know the AC source voltage in the circuit shown in Fig. 7-6, but we know that the load has a resistance of 200 Ω. The circuit is activated for 90 min, and during this time the ammeter reads 500 mA rms. How much energy is consumed by the load?

☑ **SOLUTION 7-15**

First, convert the time to hours: 90 min = 1.5 h. Therefore, $t = 1.5$. Then convert the current to amperes: 500 mA = 0.5 A. Therefore, $I = 0.5$. Note that $R = 200$. Plug these numbers into the third formula for energy given above:

$$W = (I_{rms})^2 Rt$$

$$= (0.5 \times 0.5 \times 200 \times 1.5) \text{Wh}$$

$$= 75 \text{ Wh}$$

Lightning

No discussion of home electrical systems would be complete without a mention of the phenomenon that gives everyone a bad scare sooner or later: *lightning*!

CAUSES

There is a constant potential difference between the earth's surface and the ionosphere. The lower atmosphere acts as an insulator, and the earth and the ionosphere together form a giant electrical storage device or a *capacitor*. The charge in this atmospheric capacitor is stupendous, and the insulating atmosphere frequently breaks down. That's when lightning occurs.

Lightning discharges tend to be concentrated in areas of precipitation, particularly in rain storms. Lightning is common in tropical storms and hurricanes, although some have a lot more than others. Lightning can also take place in heavy snow squalls and blizzards (especially mountain storms), and it can even strike, once in a while, from a partly cloudy sky while the sun is shining and no precipitation is falling. Sand storms and erupting volcanoes have also been known to produce lightning.

Whenever a potential difference builds up indefinitely between two natural points, a lightning discharge is bound to occur. Its capricious nature is the basis of legends and myths that go back to the beginnings of civilization.

THE STROKE

A lightning surge is called a *stroke*. It lasts for only a small fraction of a second, but within this time it can start fires, cause explosions, destroy electrical and electronic equipment, and electrocute people and animals.

There are four types of lightning that are common in and near thunderstorms. These are:

- Intracloud (within a single cloud)
- Cloud-to-ground
- Intercloud (between different clouds)
- Ground-to-cloud

The direction of current flow in a lightning stroke is usually considered to be the direction in which the electrons move, that is, from the negative to the positive. This is contrary to the physicist's definition of conventional current as a flow of charge carriers from positive to negative.

A stroke begins when the charge between a cloud and the ground, or between a cloud and some other part of the cloud, becomes so large that the electrons begin to advance through the air from the negative charge pole. A small current, called a *stepped leader*, finds the path of least resistance through the atmosphere, and causes some of the air molecules to become *ionized* (electrically charged). This process can take from a few milliseconds to several tenths of a second. Once the stepped leader has established the circuit by ionizing a *channel* through the atmosphere, one or more *return strokes* immediately follow, attended by massive current that can peak at over 100,000 A. This current is responsible for the destructive effects of lightning.

PERSONAL PROTECTION

Lightning kills more people every year in the United States than hurricanes or tornadoes. The main danger to people is from burns and electrocution. In the case of property, the damage can result from fires, induced currents, and explosions. People can be harmed by lightning, and electronic equipment damaged, despite precautions. However, the danger can be minimized.

Lightning can take place at any time. If thunder can be heard, lightning is taking place within a few kilometers of your location, whether it can be seen or not. The following precautions can minimize personal danger (but they do not guarantee immunity):

- Stay indoors, or inside an insulated metal enclosure such as a car, bus, or train.
- Stay away from windows.
- If it is not possible to get indoors, find a low spot on the ground, such as a ditch or a ravine, and squat down with your feet close together until the threat has passed.
- Avoid lone trees or other isolated, tall objects, such as utility poles or flagpoles.
- Avoid using electric appliances or electronic equipment that makes use of the utility power lines, or that has an outdoor antenna.
- Stay out of the shower or bathtub.
- Avoid swimming pools, either indoors or outdoors.
- Do not use hard-wired telephone sets.

PROTECTING HARDWARE

Precautions that minimize the risk of damage to electrical and electronic equipment (but do not guarantee immunity) are as follows:

- Never operate, or experiment with, a hobby radio station when lightning is occurring near your location. Disconnect all antennas and ground them.
- Unplug all sensitive appliances from wall outlets if a heavy thunderstorm approaches.
- Devices called *lightning arrestors* provide some protection from electrostatic-charge buildup, but they cannot offer complete safety, and should not be relied upon for routine protection.
- One or more *lightning rods* can reduce (but not eliminate) the chance of a direct hit, but they should not be used as an excuse to neglect the other precautions.
- Power line *transient suppressors* (also called "surge protectors") reduce computer "glitches" and can sometimes protect sensitive components, but again, they should not be used as an excuse to neglect the other precautions.
- Other secondary protection devices are advertised in electronics-related magazines.

For more information, consult a competent engineer or electrician. If you are concerned about the fire safety level of your house or business, consult your local fire inspector.

Quiz

This is an "open book" quiz. You may refer to the text in this chapter. A good score is 8 correct answers. Answers are in Appendix 1.

1. In an AC sine wave, 12.5% of a cycle is equivalent to

    (a) 12.5° of phase.
    (b) 25° of phase.
    (c) 45° of phase.
    (d) 90° of phase.

2. Consider a circuit with 3-phase AC. Call the waves in this circuit "wave 1," "wave 2," and "wave 3." The phase difference between waves 1 and 2 is

(a)   the same as the phase difference between waves 2 and 3.
(b)   half the phase difference between waves 2 and 3.
(c)   twice the phase difference between waves 2 and 3.
(d)   impossible to determine without more information.

3.   A transformer is supplied with AC electricity. The transformer is connected to a load. The voltage across the primary winding is 120 V rms,
and the primary draws 1 A rms. The voltage across the secondary is
12 V rms, and the load draws 8 A rms. What is the efficiency of this
transformer? Assume that no power is dissipated anywhere except in
the load and the transformer.

(a)   80%
(b)   96%
(c)   104%
(d)   125%

4.   Suppose an AC transformer has 100 turns in the primary winding and
200 turns in the secondary winding. This is

(a)   a step-up transformer.
(b)   a step-down transformer.
(c)   a transformer with a turns ratio of $T = 100{:}1$.
(d)   a transformer with a turns ratio of $T = 200{:}1$.

5.   Suppose an AC transformer has 100 turns in the primary winding and
200 turns in the secondary winding. If an AC sine wave with a voltage
of 117 V rms is applied to the primary winding, what is the peak-to-
peak AC voltage across the secondary winding?

(a)   58.5 V pk–pk
(b)   165 V pk–pk
(c)   234 V pk–pk
(d)   662 V pk–pk

6.   Examine Fig. 7-7. Suppose the switch is closed. The AC ammeter indicates 3 A rms, while the AC voltmeter indicates 12 V rms. What is the
load resistance? Assume that the timer has no resistance.

(a)   It depends on the timer reading.
(b)   4 Ω
(c)   36 Ω
(d)   108 Ω

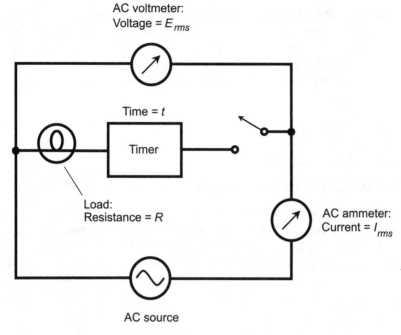

AC voltmeter:
Voltage = E_{rms}

Time = t

Timer

Load:
Resistance = R

AC ammeter:
Current = I_{rms}

AC source

Fig. 7-7.   Illustration for Quiz Questions 6 through 9.

7. Suppose the switch is closed in the circuit of Fig. 7-7. The AC ammeter indicates 3 A rms, while the AC voltmeter indicates 12 V rms. What is the power drawn by the load? Assume the timer has no resistance.

    (a) 4 W
    (b) 36 W
    (c) 108 W
    (d) More information is needed to answer this question.

8. Suppose the switch is closed in the circuit of Fig. 7-7. The AC ammeter indicates 3 A rms, while the AC voltmeter indicates 12 V rms. What is the energy consumed by the load? Assume the timer has no resistance.

    (a) 4 Wh
    (b) 36 Wh
    (c) 108 Wh
    (d) More information is needed to answer this question.

9. Suppose the switch is closed in the circuit of Fig. 7-7. The AC ammeter indicates 3 A rms, while the AC voltmeter indicates 12 V rms. The switch is opened after the timer reading has gone from an initial value of 00:00 (0 h, 0 min) to 01:45. What is the energy consumed by the load? Assume that the timer has no resistance.

(a) 4 Wh
(b) 36 Wh
(c) 108 Wh
(d) None of the above

10. All of the following can cause losses in a transformer, except

(a) resistance in the wire that makes up the windings.
(b) eddy currents in the material that makes up the core.
(c) lack of a load connected to the transformer primary.
(d) hysteresis in the material that makes up the core.

CHAPTER 8

Power Supplies

A *power supply* converts utility AC to DC for use with certain electrical and electronic devices. In this chapter, we'll examine the components of a typical power supply. Several new schematic symbols will come up. Each time a symbol appears that we haven't yet used in this book, that symbol is labeled.

- **Warning: Electrical power supplies can be dangerous. They can retain deadly voltages even after they have been switched off and unplugged. If you have any doubt about your ability to safely build or work with a power supply, leave it to a professional.**

Rectifiers

A *rectifier* converts AC to *pulsating DC*. This is usually done by means of one or more heavy-duty *semiconductor diodes*, following a power transformer. Diodes let current flow in one direction but not in the other direction.

ONE-WAY CURRENT

In a properly operating rectifier diode, the conventional current (which flows from positive to negative) can normally go only in the direction the arrow in the diode symbol points. That means the electrons, which move from negative to positive, can flow only against the arrow.

HALF-WAVE CIRCUIT

The simplest rectifier circuit, called the *half-wave rectifier* (Fig. 8-1A), has a single diode that "chops off" half of the AC cycle. The effective (eff) output voltage from a power supply that uses a half-wave rectifier is considerably less than the peak transformer output voltage, as shown in Fig. 8-2A. It is about 45% of the rms AC voltage that appears across the transformer secondary. The peak voltage in the reverse direction, called the *peak inverse voltage* (PIV) or the *peak reverse voltage* (PRV) across the diode, can be as much as 2.8 times the applied rms AC voltage.

Most engineers like to use diodes whose PIV ratings are at least 1.5 times the maximum expected PIV. Therefore, in a half-wave rectifier circuit, the diodes should be rated for at least 2.8 × 1.5, or 4.2, times the rms AC voltage that appears across the secondary winding of the power transformer.

Half-wave rectification has certain shortcomings. First, the output is difficult to filter. Second, the output voltage can drop considerably when the supply is required to deliver high current. Third, half-wave rectification puts a strain on the transformer and diodes because it *pumps* them. That means that the circuit works the diodes hard during half the AC cycle, and lets them "loaf" during the other half.

Half-wave rectification is usually adequate for use in a power supply that is not required to deliver much current, or when the voltage can vary considerably without affecting the behavior of the equipment connected to it. The main advantage of a half-wave circuit is that it costs less than more sophisticated circuits.

FULL-WAVE CENTER-TAP CIRCUIT

A better alternative for changing AC to DC takes advantage of both halves of the AC cycle. A *full-wave center-tap rectifier* has a transformer with a tapped secondary (Fig. 8-1B). The center tap is connected to an *electrical ground*, also called the *chassis ground*. This produces voltages and currents at the ends of the winding that are in phase opposition with respect to each other. These two AC waves can be individually half-wave rectified, cutting off one half of the cycle and then the other, over and over.

The effective output voltage from a power supply that uses a full-wave center-tap rectifier is slightly less than the peak transformer output voltage,

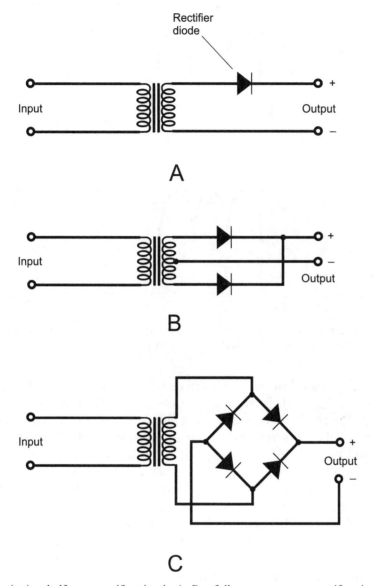

Fig. 8-1.   At A, a half-wave rectifier circuit. At B, a full-wave center-tap rectifier circuit. At C,
a full-wave bridge rectifier circuit.

as shown in Fig. 8-2B. It is about 90% of the rms AC voltage that appears across either half of the transformer secondary. The PIV across the diodes can be as much as 2.8 times the applied rms AC voltage. Therefore, the diodes should have a PIV rating of at least 4.2 times the applied rms AC voltage to ensure that they won't break down.

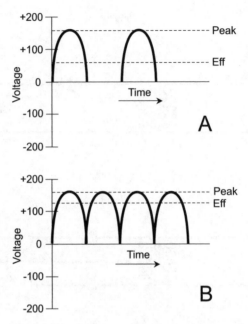

Fig. 8-2.   At A, the output of a half-wave rectifier. At B, the output of a full-wave rectifier. Note the difference in how the effective (eff) voltages compare with the peak voltages.

The output of a full-wave center-tap rectifier is easier to filter than that of a half-wave rectifier. This is because the frequency of the pulsations in the DC (known as the *ripple frequency*) from a full-wave rectifier is twice the ripple frequency of the pulsating DC from a half-wave rectifier, assuming that there is an identical AC input frequency in either situation. If you compare Fig. 8-2B with Fig. 8-2A, you should be able to see that the full-wave-rectifier output is "closer to pure DC" than the half-wave rectifier output. Another advantage of a full-wave center-tap rectifier is the fact that it's gentler with the transformer and the diodes than a half-wave rectifier. Yet another asset: With a load at the output of a power supply that uses a full-wave center-tap rectifier circuit, the voltage drops less than is the case with a half-wave supply. But because the transformer is more sophisticated, the full-wave center-tap circuit costs more than a half-wave circuit that delivers the same output voltage at the same rated maximum current.

FULL-WAVE BRIDGE CIRCUIT

Another way to get full-wave rectification is the *full-wave bridge rectifier*, often called simply a *bridge*. It is diagrammed in Fig. 8-1C. The output waveform is similar to that of the full-wave center-tap circuit (Fig. 8-2B).

The effective output voltage from a power supply that uses a full-wave bridge rectifier is slightly less than the peak transformer output voltage, as shown in Fig. 8-2B. It is about 90% of the rms AC voltage that appears across the transformer secondary. The PIV across the diodes in the bridge circuit is about 1.4 times the applied rms AC voltage. Therefore, each diode needs to have a PIV rating of at least 1.4×1.5, or 2.1, times the rms AC voltage that appears at the transformer secondary.

The bridge circuit does not require a center-tapped transformer secondary. It uses the entire secondary winding on both halves of the wave cycle, so it makes even more efficient use of the transformer than the full-wave center-tap circuit. The bridge is also easier on the diodes than half-wave or full-wave center-tap circuits.

VOLTAGE-DOUBLER CIRCUIT

Diodes and *capacitors* (components that hold electric charge) can be interconnected to deliver a DC output that is approximately twice the positive or negative peak AC input voltage. In practice, a *voltage-doubler power supply* works well only when the load draws a low current. Otherwise, the *voltage regulation* is poor; the voltage drops a lot when the current demand is significant. The best way to build a high-voltage power supply is to use a step-up transformer, not a voltage-doubling scheme. Nevertheless, a voltage-doubler supply can be, and sometimes is, used when the cost of the circuit must be minimized and the demands placed on it are expected to be modest.

Figure 8-3 is a simplified diagram of a voltage-doubler power supply. It works on the entire AC cycle, and so it is called a *full-wave voltage doubler*. This circuit subjects the diodes to a PIV of 2.8 times the applied rms AC voltage. Therefore, they should be rated for PIV of at least 4.2 times the rms AC voltage that appears across the transformer secondary.

Proper operation of this type of circuit depends on the ability of the capacitors to hold a charge under maximum load. This means that the capacitors must have large values, as well as being capable of handling high voltages. The capacitors serve two purposes: to boost the voltage and to filter the output. The resistors, which have low ohmic values, protect the diode against *surge currents* that occur when the power supply is first switched on.

[?] **PROBLEM 8-1**

Suppose a power transformer with a 1:2 primary-to-secondary turns ratio is connected to the 117 V AC utility mains. A half-wave rectifier circuit is used

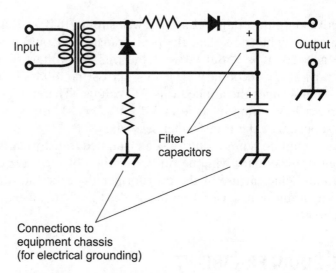

Fig. 8-3. A full-wave voltage-doubler power supply.

to obtain pulsating DC output. What is the minimum PIV rating the diodes should have in order to ensure that they won't break down? Round the answer off to the nearest volt.

✓ SOLUTION 8-1

The utility mains supply 117 V rms AC, and the transformer steps up this voltage by a factor equal to the primary-to-secondary turns ratio. Therefore, the output of the transformer is 2×117 V rms AC, or 234 V rms AC. In a half wave circuit, the PIV rating of the diodes should be at least 4.2 times this value. That means they must be rated for at least 4.2×234 PIV, or 983 PIV.

? PROBLEM 8-2

Suppose a full-wave bridge rectifier circuit is used in the above situation, instead of a half-wave rectifier circuit. What must be the minimum PIV rating of the diodes in this case? Round the answer off to the nearest volt.

✓ SOLUTION 8-2

The secondary still delivers 234 V rms AC. In a full-wave bridge circuit, the PIV rating of the diodes should be at least 2.1 times the rms AC voltage at the transformer secondary. This is half the PIV in the half-wave situation: 2.1×234 PIV, or 491 PIV.

PROBLEM 8-3

Suppose a voltage-doubler circuit is used in the above scenario. What must be the minimum PIV rating of the diodes in this case? Round the answer off to the nearest volt.

✔ **SOLUTION 8-3**

The transformer secondary still delivers 234 V rms AC. In a voltage-doubler circuit, the PIV rating of the diodes should be at least 4.2 times the rms AC voltage at the transformer secondary. This is the same as the PIV in the half-wave situation: 4.2×234 PIV, or 983 PIV.

Filtering and Regulation

Most DC-powered devices need something better than the pulsating DC that comes right out of a rectifier circuit. The pulsations (ripple) in the rectifier output can be eliminated by a *filter*.

CAPACITORS ALONE

The simplest power-supply filter consists of one or more large-value capacitors, connected in parallel with the rectifier output (Fig. 8-4). A good component for this purpose is known as an *electrolytic capacitor*. This type of capacitor is *polarized*, meaning that it must be connected in the correct direction in the circuit. Each unit is also rated for a certain

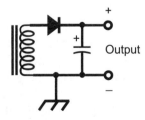

Fig. 8-4. A large-value capacitor can be used all by itself as a power-supply filter.

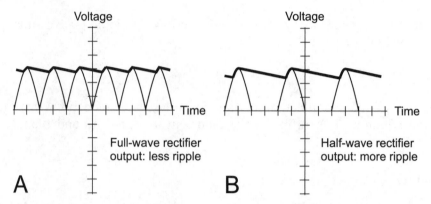

Fig. 8-5. Filtering of ripple from a full-wave rectifier (A) and from a half-wave rectifier (B).

maximum voltage. Pay attention to these ratings if you ever have occasion to work with "electrolytics"!

Filter capacitors work by "trying" to maintain the DC voltage at its peak level, as shown in Fig. 8-5. This is easier to do with the output of a full-wave rectifier (drawing A) than with the output of a half-wave rectifier (drawing B). With a full-wave rectifier receiving a 60 Hz AC electrical input, the ripple frequency is 120 Hz; with a half-wave rectifier, it is 60 Hz. The filter capacitors are recharged twice as often with a full-wave rectifier, as compared with a half-wave rectifier. This keeps the output DC more "pure."

CAPACITORS AND CHOKES

Another way to smooth out the DC from a rectifier is to place a large-value *inductor* in series with the output, and a large-value capacitor in parallel. The inductor is called a *filter choke*.

In a filter that uses a capacitor and an inductor, the capacitor can be placed on the rectifier side of the choke. This is a *capacitor-input filter* (Fig. 8-6A). If the filter choke is placed on the rectifier side of the capacitor, the circuit is a *choke-input filter* (Fig. 8-6B). Capacitor-input filtering can be used when a power supply is not required to deliver much current. The output voltage is higher with a capacitor-input filter than with a choke-input filter having identical inputs. If the supply needs to deliver large or variable amounts of current, a choke-input filter is a better choice, because the output voltage is more stable.

If the output of a power supply must have an absolute minimum of ripple, two or three capacitor/choke pairs can be connected in *cascade* (Fig. 8-7).

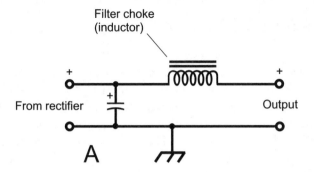

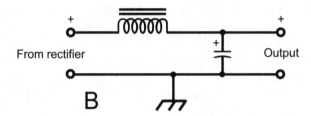

Fig. 8-6. At A, a capacitor-input filter. At B, a choke-input filter.

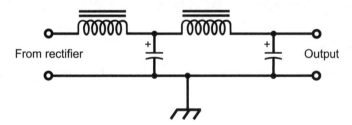

Fig. 8-7. Two choke-input filter sections in cascade.

Each pair constitutes a *section* of the filter. Multi-section filters can consist of capacitor-input or choke-input sections, but the two types are never mixed.

In the example of Fig. 8-7, both capacitor/choke pairs are called *L sections*. If the second capacitor is omitted, the filter becomes a *T section*. If the second capacitor is moved to the input and the second choke is omitted, the filter becomes a *pi section*. These sections are named because their schematic diagrams look something like the uppercase English L, the uppercase English T, and the uppercase Greek Π, respectively.

VOLTAGE REGULATION

If a special diode called a *Zener diode* is connected in parallel with the output of a power supply, the diode limits the output voltage. The diode must have an adequate power rating to prevent it from burning out. The limiting voltage depends on the particular Zener diode used. There are Zener diodes available that will fit any reasonable power-supply voltage. Figure 8-8 is a diagram of a full-wave bridge DC power supply including a Zener diode for voltage regulation. Note the direction in which the Zener diode is connected in this application: with the arrow pointing from minus to plus. This is contrary to the polarity used for rectifier diodes. It's important that the polarity be correct with a Zener diode, or it will burn out!

A Zener-diode voltage regulator is inefficient when the supply is used with equipment that draws high current. When a supply must deliver a lot of current, a *power transistor* is used along with the Zener diode to obtain regulation. A circuit diagram of such a scheme is shown in Fig. 8-9.

Voltage regulators are available in *integrated-circuit* (IC) form. The *regulator IC*, also called a *regulator chip*, is installed in the power-supply circuit at the output of the filter. In high-voltage power supplies, *electron-tubes* are sometimes used as voltage regulators. These are particularly rugged and can withstand much higher temporary overloads than the Zener diodes, transistors, or chips. However, *regulator tubes* are considered old-fashioned—even archaic—by some engineers.

? PROBLEM 8-4

Suppose the standard AC line frequency is higher than 60 Hz. Does this make the output of a rectifier circuit easier to filter, or more difficult?

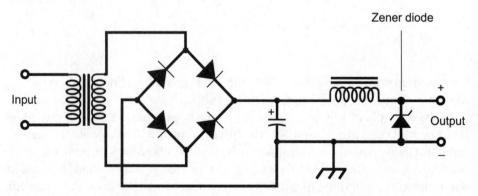

Fig. 8-8. A power supply with a Zener-diode voltage regulator in the output.

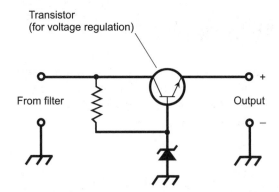

Transistor
(for voltage regulation)

From filter

Output

+

−

Fig. 8-9.   A voltage-regulator circuit using a Zener diode and an NPN transistor.

✔ **SOLUTION 8-4**

In theory, it makes the output of a rectifier circuit easier to filter, because the capacitors don't have to hold the charge for so long.

? **PROBLEM 8-5**

How can a technician tell when an electrolytic capacitor is connected with the polarity correct?

✔ **SOLUTION 8-5**

An electrolytic capacitor is labeled with either a plus sign or a minus sign. The wires (called *leads*) or lugs should be connected in the circuit so that the plus (+) side of the capacitor goes to the part of the circuit with the more positive voltage, or so the (−) side goes to the part of the circuit with the more negative voltage.

? **PROBLEM 8-6**

What will happen if an electrolytic capacitor is connected with the polarity reversed, or to a circuit with a voltage higher than its rated voltage?

✔ **SOLUTION 8-6**

It won't provide the necessary capacitance. If the voltage is excessive, the component may explode.

Protection of Equipment

The output of a power supply should be free of any sudden changes that can damage equipment or components, or interfere with their proper performance. It is also important that voltages do not appear on the external surfaces of a power supply, or on the external surfaces of any equipment connected to it.

GROUNDING

The best electrical ground for a power supply is the "third wire" ground provided in up-to-date AC utility circuits. In an AC outlet, this connection appears as a "hole" shaped like an uppercase letter D turned on its side. The contacts inside this "hole" should be connected to a wire that ultimately terminates in a metal rod driven into the earth at the point where the electrical wiring enters the building. That constitutes an *earth ground*.

In older buildings, *2-wire AC systems* are common. These can be recognized by the presence of two slots in the utility outlets, but no ground "hole." Some of these systems employ reasonable grounding by means of a scheme called *polarization*, where one slot is longer than the other, the longer slot being connected to an electrical ground. But this is not as good as a *3-wire AC system*, where the ground connection is independent of both the outlet slots.

Unfortunately, the presence of a three-wire or a polarized outlet system does not always mean that an appliance connected to an outlet is well-grounded. If the appliance design is faulty, or if the ground "holes" at the outlets were not grounded by the people who installed the electrical system, a power supply can deliver unwanted voltages to the external surfaces of appliances and electronic devices. This can present an electrocution hazard, and can also hinder the performance of the equipment connected to the supply.

- **Warning: All exposed metal surfaces of power supplies should be connected to the grounded wire of a three-wire electrical cord. The "third prong" of the plug should never be defeated or cut off. Some means should be found to ensure that the electrical system in the building has been properly installed, so you don't work under the illusion that your system has a good ground when it really doesn't. If you are in doubt about this, consult a professional electrician.**

SURGE CURRENTS

At the instant a power supply is switched on, a surge of current occurs, even with nothing connected to the supply output. This is because the filter capacitors need an initial charge, so they draw a large current for a short time. The *surge current* is far greater than the normal operating current. An extreme surge can destroy the rectifier diodes in a power supply. The phenomenon is worst in high-voltage supplies and voltage-multiplier circuits. A diode failure, as a result of current surges, can be prevented in at least three ways:

- Use diodes with a current rating of many times the normal operating level.
- Connect several diodes in parallel wherever a diode is needed in the circuit. *Current-equalizing resistors* are necessary (Fig. 8-10). The resistors should have small ohmic values. All the diodes should be identical.
- Use an *automatic switching* circuit in the transformer primary. This type of circuit applies a reduced AC voltage to the transformer for 1 or 2 seconds, and then applies the full input voltage.

TRANSIENTS

The AC on the utility line is a sine wave with a constant voltage near 117 V rms or 234 V rms. But there are often *voltage spikes*, known as *transients*, that can attain positive or negative peak values of several thousand volts. Transients are caused by sudden changes in the load in a utility circuit. A heavy thundershower can produce transients throughout an entire neighborhood or a small town. Unless they are suppressed, transients can destroy the diodes in a power supply. Transients can also befuddle sensitive electronic equipment, such as computers or microcomputer-controlled appliances.

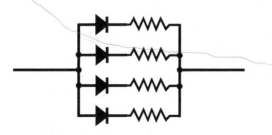

Fig. 8-10.   Diodes in parallel, with current-equalizing resistors in series with each diode.

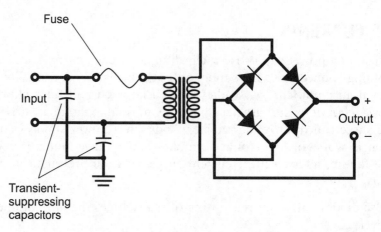

Fig. 8-11. A full-wave bridge rectifier with transient-suppression capacitors and a fuse in the transformer primary circuit.

The simplest way to get rid of common transients is to place a small capacitor of about $0.01\,\mu F$, rated for 600 V or more, between each side of the transformer primary and electrical ground, as shown in Fig. 8-11. A good component for this purpose is a *disk ceramic capacitor* (not an electrolytic capacitor). Disk ceramic capacitors have no polarity issues. They can be connected in either direction work equally well.

Commercially made *transient suppressors* are available. These devices, often mistakenly called "surge protectors," use more sophisticated methods to prevent sudden voltage spikes from reaching levels where they can cause problems. It is a good idea to use transient suppressors with all sensitive electronic devices, including computers, hi-fi stereo systems, and television (TV) sets. In the event of a thundershower, the best way to protect such equipment is to physically unplug it from the wall outlets until the event has passed.

FUSES

A *fuse* is a piece of soft wire that melts, breaking a circuit if the current exceeds a certain level. A fuse is placed in series with the transformer primary, as shown in Fig. 8-11. A short circuit or overload anywhere in the power supply, or in equipment connected to it, will burn the fuse out. If a fuse blows out, it must be replaced with another one of the same rating. Fuses are rated in amperes (A). Thus, a 5 A fuse will carry up to 5 A for a little while before blowing out, and a 20 A fuse will carry up to 20 A.

Fuses are available in two types: the *quick-break fuse* and the *slow-blow fuse*. A quick-break fuse is a straight length of wire or a metal strip. A slow-blow fuse usually has a spring inside along with the wire or strip. It's best to replace blown-out fuses with new ones of the same type. Quick-break fuses in slow-blow situations can burn out needlessly, causing inconvenience. Slow-blow fuses in quick-break environments might not provide adequate protection to the equipment, letting excessive current flow for too long before blowing out.

CIRCUIT BREAKERS

A *circuit breaker* performs the same function as a fuse, except that a breaker can be reset by turning off the power supply, waiting a moment, and then pressing a button or flipping a switch. Some breakers reset automatically when the equipment has been shut off for a certain length of time. Circuit breakers are rated in amperes (A), just like fuses.

If a fuse or breaker keeps blowing out or tripping, or if it blows or trips immediately after it has been replaced or reset, then something is wrong with the power supply or with the equipment connected to it. Burned-out diodes, a bad transformer, and shorted filter capacitors in the supply can all cause this trouble. A short circuit in the equipment connected to the supply, or the connection of a device in the wrong direction (polarity), can cause repeated fuse blowing or circuit-breaker tripping.

Never replace a fuse or a breaker with a larger-capacity unit to overcome the inconvenience of repeated fuse/breaker blowing/tripping. Find the cause of the trouble, and repair the equipment accordingly. The "penny in the fuse box" scheme can endanger equipment and personnel, and it increases the risk of fire in the event of a short circuit.

THE FINISHED SYSTEM

Figure 8-12 is a block diagram of a complete power supply. Note the order in which the portions of the system, called *stages*, are connected.

PROBLEM 8-7

Will a transient suppressor work properly if it is designed for a three-wire electrical system, but the ground wire has been defeated, cut off, or does not lead to a good electrical ground?

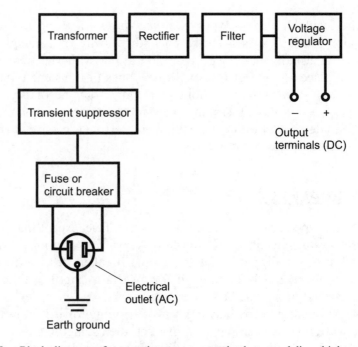

Fig. 8-12. Block diagram of a complete power supply that can deliver high-quality DC
output with AC input.

☑ **SOLUTION 8-7**

No. In order to function properly, a transient suppressor needs a substantial
ground. The excessive, sudden voltages are shunted away from sensitive
equipment only when a current path is provided to allow discharge to
ground.

? **PROBLEM 8-8**

What is the difference between a surge and a transient?

☑ **SOLUTION 8-8**

The term *surge* refers to the initial high current drawn by a cheap or poorly
designed power supply when it's first switched on with a load connected. The
term *transient* refers to a high-voltage "spike" that can be induced in the AC
line by lightning, sparking (known as *arcing*) in utility power transformers, or
arcing in equipment connected in the same utility circuit as the supply.

? **PROBLEM 8-9**

Isn't it extreme to recommend that sensitive equipment be physically unplugged from wall outlets during thundershowers? Isn't a switch on the power supply, or in the house wiring, good enough?

✔ **SOLUTION 8-9**

Unplugging things is inconvenient, but it is the only way to absolutely ensure protection. A single lightning strike on a power line near your house or business can induce a voltage high enough to jump a gap of several centimeters. The current induced by such an electrical potential can arc across an open switch easily!

Quiz

This is an "open book" quiz. You may refer to the text in this chapter. A good score is 8 correct answers. Answers are in Appendix 1.

1. Examine the half-wave rectifier circuit shown in Fig. 8-1A. Suppose the diode were connected in the opposite direction from that shown. What would happen to the output?

    (a)   Nothing. The output would stay exactly the same.
    (b)   The output polarity would be reversed, and the voltage would be different.
    (c)   The output polarity would be reversed, but the voltage would be the same.
    (d)   The output polarity would be the same, but the voltage would be different.

2. A capacitor-input filter is usually satisfactory

    (a)   for use between the transformer and the rectifier diodes in a power supply.
    (b)   for use in the primary circuit of a transformer.
    (c)   in power supplies connected to equipment that demands excellent voltage regulation.
    (d)   in power supplies connected to loads that don't draw much current.

3. Examine Fig. 8-2. Assume that the AC frequency at the power supply transformer primary is 60 Hz in the situations portrayed by both graphs, A and B. If that is true, then the ripple frequency of the waveform at B is

    (a) twice the ripple frequency of the waveform at A.
    (b) half the ripple frequency of the waveform at A.
    (c) the same as the frequency of the waveform at A.
    (d) impossible to determine without more information.

4. Examine Fig. 8-2. Assume the AC voltage across the secondary winding of the transformer is 250 V rms in the situations portrayed by both graphs, A and B. If that is true, then the positive peak voltage of the waveform at B is

    (a) twice the positive peak voltage of the waveform at A.
    (b) half the positive peak voltage of the waveform at A.
    (c) the same as the positive peak voltage of the waveform at A.
    (d) impossible to determine without more information.

5. The effective output voltage from a half-wave rectifier, compared with the rms voltage across the transformer primary winding, depends on

    (a) the PIV rating of the rectifier diode.
    (b) the turns ratio of the transformer.
    (c) the direction in which the diode is connected.
    (d) All of the above

6. Suppose a power transformer with a 3:1 primary-to-secondary turns ratio is connected to the 117 V AC utility mains. A half-wave rectifier circuit is used to obtain pulsating DC output. What is the minimum PIV rating the diodes should have in order to ensure that they won't break down? Round the answer off to the nearest volt.

    (a) 82 PIV
    (b) 164 PIV
    (c) 737 PIV
    (d) 1474 PIV

7. Examine Fig. 8-4. What would happen to the output if the capacitor were removed?

    (a) The output voltage would drop to zero.
    (b) The output voltage would increase.
    (c) The output would change from filtered DC to pulsating DC.
    (d) The output would change from DC to AC.

8. Examine Fig. 8-4. What would happen if the diode were to blow out, resulting in an open circuit in its place?

   (a) The output voltage would drop to zero.
   (b) The output voltage would increase.
   (c) The output would change from filtered DC to pulsating DC.
   (d) The output would change from DC to AC.

9. Examine Fig. 8-4. Suppose you decide you want to reverse the polarity of the output of this supply. Assume the capacitor is of the electrolytic type. What should you do?

   (a) Turn the transformer around, so the primary winding becomes the secondary, and the secondary winding becomes the primary.
   (b) Reverse the polarity of the diode.
   (c) Reverse the polarity of the capacitor.
   (d) Reverse the polarities of both the diode and the capacitor.

10. A multistage cascaded L-network power-supply filter consists of

   (a) inductors in series with the rectifier output, and capacitors in parallel with it.
   (b) inductors in parallel with the output, and capacitors in series with it.
   (c) inductors in series with the transformer primary, and capacitors across it.
   (d) inductors across the transformer primary, and capacitors in series with it.

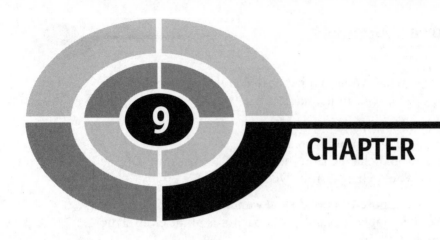

CHAPTER

Wire and Cable

Wire is the most common medium for transferring electrical energy from one place to another. Some types of wire are used to make lamps, resistors, and heating elements. A *cable* is a special cord that is designed to carry electric currents or signals.

Wire Conductors

Silver is the best known electrical conductor, followed by copper and aluminum. Physically, steel is the strongest material available at a moderate cost. Copper and aluminum, while better electrical conductors than steel, are physically weaker.

SOLID VS STRANDED

Some wire consists of a single, solid, round conductor. This is called *solid wire*. Some wire has several uninsulated, thin, round conductors bundled or woven together to form a single, larger round conductor. This is called *stranded wire*.

GAUGE

The diameter of a wire is defined by a number called the *gauge*. In general, higher gauge numbers correspond with smaller diameters. There are three different gauge scales in common use today.

The *American Wire Gauge* (*AWG*) is expressed as a whole number. Table 9-1 shows wire diameters, in millimeters (mm) and inches (in.), for AWG No. 1 through AWG No. 40. These figures are for the metal only, and do not include any insulation or enamel.

In some countries, *British Standard Wire Gauge* (*NBS SWG*) is used. The British Standard Wire Gauge sizes for designators 1 through 40 are shown in Table 9-2. The NBS SWG figures, like those for AWG, do not take into account any coatings on the wire.

The *Birmingham Wire Gauge* (*BWG*) designators differ from the American and the British Standard designators, but the sizes are nearly the same. Table 9-3 shows the diameter vs BWG for designators 1 through 20. The BWG designator, like the others, does not include any coatings or insulation.

RESISTIVITY

No wire is a perfect conductor. All wire offers some resistance to the flow of current. This opposition is expressed in terms of its *resistivity*, or resistance per unit length. A common unit is the *micro-ohm per meter* ($\mu\Omega$/m).

For a given metal, larger gauges (smaller diameters) of wire have greater resistivity than smaller gauges (larger diameters). In the case of a solid copper wire carrying DC at room temperature, approximate resistivity in $\mu\Omega$/m for an even-numbered solid-copper wire sizes from AWG No. 2 through No. 30 are shown in Table 9-4.

CARRYING CAPACITY

The ability of a wire to handle DC and utility AC safely is called its *carrying capacity*. This specification is usually given in amperes (A). Table 9-5 shows an approximate DC and low-frequency rms AC carrying capacity for an even-numbered solid-copper wire sizes from AWG No. 8 through No. 20, in open air at room temperature.

Wire can intermittently carry larger currents than those shown in the table, but the danger of softening or melting, with consequent breakage, rises rapidly as the current increases beyond these values. The fire hazard also increases when the wire gets too hot.

Table 9-1   American Wire Gauge (AWG) diameters.

AWG	Millimeters	Inches
1	7.35	0.289
2	6.54	0.257
3	5.83	0.230
4	5.19	0.204
5	4.62	0.182
6	4.12	0.163
7	3.67	0.144
8	3.26	0.128
9	2.91	0.115
10	2.59	0.102
11	2.31	0.0909
12	2.05	0.0807
13	1.83	0.0720
14	1.63	0.0642
15	1.45	0.0571
16	1.29	0.0508
17	1.15	0.0453
18	1.02	0.0402
19	0.912	0.0359
20	0.812	0.0320

Table 9-1   American Wire Gauge (AWG) diameters (*Continued*).

AWG	Millimeters	Inches
21	0.723	0.0285
22	0.644	0.0254
23	0.573	0.0226
24	0.511	0.0201
25	0.455	0.0179
26	0.405	0.0159
27	0.361	0.0142
28	0.321	0.0126
29	0.286	0.0113
30	0.255	0.0100
31	0.227	0.00894
32	0.202	0.00795
33	0.180	0.00709
34	0.160	0.00630
35	0.143	0.00563
36	0.127	0.00500
37	0.113	0.00445
38	0.101	0.00398
39	0.090	0.00354
40	0.080	0.00315

Table 9-2 British Standard Wire Gauge (NBS SWG) diameters.

NBS SWG	Millimeters	Inches
1	7.62	0.300
2	7.01	0.276
3	6.40	0.252
4	5.89	0.232
5	5.38	0.212
6	4.88	0.192
7	4.47	0.176
8	4.06	0.160
9	3.66	0.144
10	3.25	0.128
11	2.95	0.116
12	2.64	0.104
13	2.34	0.092
14	2.03	0.080
15	1.83	0.072
16	1.63	0.064
17	1.42	0.056
18	1.22	0.048
19	1.02	0.040
20	0.91	0.036

Table 9-2   Continued.

NBS SWG	Millimeters	Inches
21	0.81	0.032
22	0.71	0.028
23	0.61	0.024
24	0.56	0.022
25	0.51	0.020
26	0.46	0.018
27	0.42	0.0164
28	0.38	0.0148
29	0.345	0.0136
30	0.315	0.0124
31	0.295	0.0116
32	0.274	0.0108
33	0.254	0.0100
34	0.234	0.0092
35	0.213	0.0084
36	0.193	0.0076
37	0.173	0.0068
38	0.152	0.0060
39	0.132	0.0052
40	0.122	0.0048

Table 9-3　Birmingham Wire Gauge (BWG) diameters.

BWG	Millimeters	Inches
1	7.62	0.300
2	7.21	0.284
3	6.58	0.259
4	6.05	0.238
5	5.59	0.220
6	5.16	0.203
7	4.57	0.180
8	4.19	0.165
9	3.76	0.148
10	3.40	0.134
11	3.05	0.120
12	2.77	0.109
13	2.41	0.095
14	2.11	0.083
15	1.83	0.072
16	1.65	0.064
17	1.47	0.058
18	1.25	0.049
19	1.07	0.042
20	0.889	0.035

Table 9-4   Resistivity of various gauges of a solid copper wire, in micro-ohms per meter, to three significant figures.

Wire size, AWG	μΩ/m
2	523
4	831
6	1320
8	2100
10	3340
12	5320
14	8450
16	13,400
18	21,400
20	34,000
22	54,000
24	85,900
26	137,000
28	217,000
30	345,000

When the wires are run alongside other electronic components, the figures in the table should be reduced somewhat. The same is true when wires are bundled into cables, and/or when wires are run near flammable materials or surrounded by insulation.

? PROBLEM 9-1

How does the resistivity of solid wire compare with wire diameter in linear units, such as millimeters?

Table 9-5 Maximum safe continuous DC carrying capacity, in amperes, for various wire sizes (AWG) in open air.

Wire size, AWG	Current, A
8	73
10	55
12	41
14	32
16	22
18	16
20	11

✔ **SOLUTION 9-1**

The resistivity of a solid wire is inversely proportional to its cross-sectional area. The cross-sectional area is the surface area of the flat circular region that appears at the end of a solid wire that is cut straight across. This area varies in direct proportion to the square of the diameter. Therefore, the resistivity decreases in proportion to the square of the diameter. If the diameter is halved, the resistivity is quadrupled. If the diameter is tripled, the resistivity decreases by a factor of 9.

? **PROBLEM 9-2**

A solid wire has a slightly lower resistivity than a stranded wire of the same gauge. Why?

✔ **SOLUTION 9-2**

Imagine a piece of solid wire cut straight across. The cross section is all metal. Now imagine a piece of stranded wire of the same outside diameter (not including any insulation) cut straight across. The cross section is not all metal; there are gaps because each strand has a disk-shaped cross section, and when the disks are squeezed together, there are gaps between them no matter how tightly they are squeezed. The "metallic cross sectional area" of a solid wire, for a given gauge, is therefore greater than the "metallic cross sectional area" of a stranded wire. This means that the solid wire conducts DC or 60-Hz utility AC a little better than the stranded wire of the same gauge. That means the resistivity of the solid wire is lower; the *conductivity* of wire is inversely proportional to the resistivity.

Wire Splicing

When wiring electrical circuits, it is often necessary to splice two lengths of wire. There are various ways to do this. Two of the most common schemes are described and illustrated here.

TWIST SPLICE

The simplest way to splice two wires is to bring the exposed ends close together and parallel. Then the ends are twisted over each other several times (Fig. 9-1). This is called a *twist splice*, and it can be used with solid wire or stranded wire. If the wires are of unequal diameters, the smaller wire is twisted around the larger wire (Fig. 9-2). An electrical tape can be put over the connection, if insulation is important.

A twist splice has poor mechanical strength, but it is convenient for temporary connections between wires. Twist splices are also common inside utility outlets and lamps, where mechanical strength is not an issue. These twist splices are covered with plastic caps to provide insulation, and to make

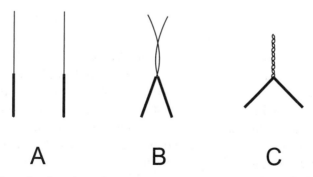

A   B   C

Fig. 9-1. Twist splice for wires of the same gauge. Wires are brought parallel (A), looped around each other (B), and then twisted over each other several times (C).

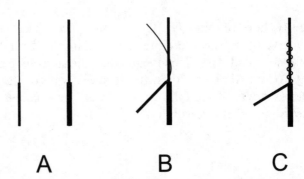

Fig. 9-2.   Twist splice for wires of differing gauge. Wires are brought parallel (A), and then the smaller wire is looped (B) and twisted (C) around the larger one.

it easy to "unsplice" the wires, if it is necessary to replace the outlet or a part of the appliance.

WESTERN UNION SPLICE

When a splice must have the maximum possible tensile strength, the wires are brought together end-to-end, overlapping about 2 in. (5 mm). The wires are hooked around each other, and then twisted several times (Fig. 9-3). This is known as a *Western Union splice*.

For guy wires or other wires intended to support weight rather than conduct electricity, splices should be avoided, if possible. If splices are necessary, the Western Union method is recommended. Each end should be twisted around 10 to 12 times. Needle-nosed pliers can be used to secure the extreme ends. The protruding ends are removed using a diagonal cutter.

SOLDERING

Splices can be soldered if the best possible electrical bond is needed. A layer of electrical tape or other insulation should be applied afterward. Soldering can also provide additional tensile strength to a splice. For large-diameter wires, the ends should be "tinned" with solder before the splice is made to optimize the electrical bond. For maximum physical strength, both the wires should be of the same size and of the same type (both solid, or both stranded). Special solder is required if aluminum wire is used, because aluminum will not adhere to ordinary solder.

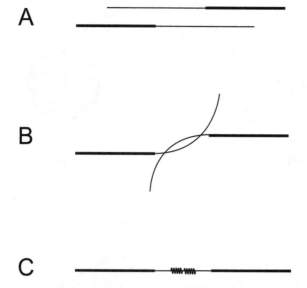

Fig. 9-3.   Western Union splice. Wires are brought parallel (A), looped around each other (B), and finally twisted over each other (C).

Electrical Cable

The simplest electrical cable, other than a plain single wire, is *lamp cord*. It is used with common appliances at low to moderate current levels. Two or three wires are embedded in rubber or plastic insulation (Fig. 9-4A). The individual conductors are stranded. This makes the conductors resistant to breakage from repeated flexing.

MULTICONDUCTOR CABLE

When a cable has several wires, they can be insulated individually, bundled together, and enclosed in an insulating jacket (Fig. 9-4B). If the cable must be flexible, each wire is stranded. Some cables of this type have dozens of conductors.

If there are only a few conductors, they can be run parallel to each other in a flat configuration, as shown in Fig. 9-4C.

Sometimes, several conductors are molded into a common insulating jacket, as shown in Fig. 9-4D. This is known as *ribbon cable*. It is used inside commercially manufactured, high-tech electronic devices, particularly

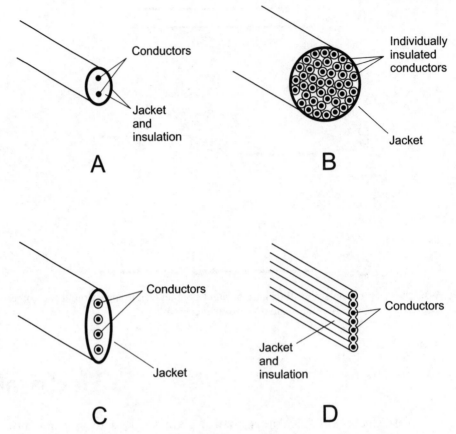

Fig. 9-4. Simple two-wire cable (A), multiconductor cable (B), flat cable (C), and ribbon cable (D).

computers. This cable is physically sturdy, is ideal for saving space, and efficiently radiates heat away.

SHIELDING

The afore-mentioned cable types are *unshielded*. For DC and utility AC, unshielded cables are usually all right. But radio signals, video, and high-speed data generate *electromagnetic fields* that can be transferred among cable conductors, and even between a cable and the surrounding environment. In these situations, an *electromagnetic shielding* is necessary.

A wire, or a group of wires, is shielded by enclosing it in a conductive cylinder of solid metal, metallic braid (usually copper), or metal foil.

This so-called *shield* is separated from the conductor or conductors by an insulating material, such as polyethylene.

In some multiconductor cables, a single shield surrounds all the wires. In other cables, each wire has its own shield. The entire cable can be surrounded by a copper braid, in addition to individual shielding of the wires. *Double-shielded cable* is surrounded by two concentric braids separated by insulation.

COAXIAL CABLE

Coaxial cable, also called *coax* (pronounced "*co*-ax"), is especially designed for high-frequency signal transmission. It is used in community-antenna television (CATV) networks and in some computer local area networks (LANs). Coaxial cable is employed by amateur and Citizens Band radio operators to connect transceivers, transmitters, and receivers to antennas. It is also used in high-end audio systems to interconnect components.

In coax, a single *center conductor* is surrounded by a cylindrical shield. In some cases, solid or foamed polyethylene insulation, called the *dielectric*, keeps the center conductor at the central axis of the cable (Fig. 9-5A). Other cables have a bare center conductor, and a thin tubular layer of polyethylene just inside the braid (drawing B), so most of the dielectric is air. Technically, this is not true coaxial cable, although it is often called by that name.

Some coaxial cables have a solid metal pipe surrounding the center conductor. This is called *hard line*. This type of cable is available in larger diameters than coaxial cables, and has lower loss per unit length. It is used in high-power, fixed radio and television transmitting installations.

In a coaxial cable, the signal is carried by the center conductor. The shield is connected to the ground to keep signals from "leaking out." The shield also keeps unwanted electromagnetic fields from getting in.

SERIAL AND PARALLEL

A *serial cable* carries current or signals along a single path from one point to another. A *parallel cable* carries current or signals along several paths from one point to another, independently and at the same time.

Serial cables can generally be longer than parallel cables, because parallel cables are prone to *crosstalk*, a condition in which signals in the different conductors interfere or combine with one another. Crosstalk can be prevented by individually shielding each conductor within the cable, but

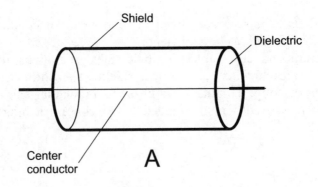

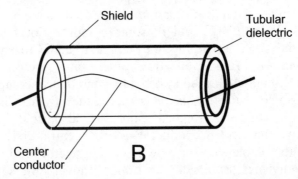

Fig. 9-5. At A, coaxial cable with solid or foamed insulation (called the dielectric) between the center conductor and the shield. At B, similar cable in which most of the dielectric is air.

this increases the physical bulk of the cable, and also increases the cost per unit length.

An example of a serial cable is the coaxial line used to carry television signals in CATV. The cord connecting the main unit and the printer in an older personal computer system is usually a parallel cable, but in a newer computer it is usually a specialized serial cable. In recent years, improved *serial data transmission* technology has made parallel cables nearly obsolete in computer systems, except for short runs between circuit boards inside the computer.

CABLE SPLICING

When lengths of multiconductor cord or cable must be spliced, it is best to use Western Union splices for each conductor. Insulation is important; all the splices should be wrapped with an electrical tape, and the entire combination wrapped afterwards. Wrapping the individual connections ensures

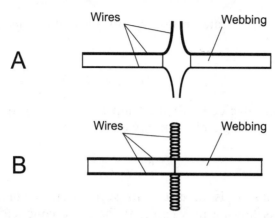

Fig. 9-6.   Twist splice for two-wire cord or cable. Ends are brought together (A) and the conductors are twisted at right angles (B). The twists are then soldered, trimmed, folded back, and insulated.

that no two conductors will come into electrical contact. Wrapping the whole combination keeps the cable from shorting out to external metallic objects, and also provides additional protection against corrosion and oxidation.

In order to minimize the risk of a short circuit between or among the conductors in a multiconductor cable, the splices for each conductor can be made at slightly different points along the cord or the cable. It is also important to ensure that each conductor is spliced to its correct counterpart! This can get confusing with multiconductor cables, especially if the color-coding scheme for the wires in one cable differs from the color-coding scheme for the wires in the other cable.

Two-conductor cord or ribbon can be twist-spliced, if necessary (Fig. 9-6). After the twists have been soldered, the splices are trimmed to about 1/2 in. (1.2 cm), folded back parallel to the cable axis, and insulated. The whole junction is then carefully wrapped with an electrical tape or coated with a sealant.

?   **PROBLEM 9-3**

Can coaxial cable be used to carry DC?

✔   **SOLUTION 9-3**

Yes, it can. In some situations, in fact, coaxial cable is preferred for DC. But coaxial cable is more expensive than the two-wire cable, such as lamp cord, with conductors of the same resistivity. In most situations, a lamp cord is good enough.

⌕ **PROBLEM 9-4**

When is a coaxial cable better than a lamp cord for carrying DC or low-frequency AC, such as single-phase utility electricity?

✔ **SOLUTION 9-4**

In some situations, it is necessary to keep electromagnetic fields away from the currents flowing in a cable. If the cable conductors are exposed to these fields, the fields can induce additional, and unwanted, current that can cause some devices to malfunction. An example is the cable used to connect the speakers of a hi-fi system to the amplifier. The use of a coaxial cable in place of a typical speaker cable (which resembles lamp cord) can eliminate, or at least reduce, the risk that a strong electromagnetic field will induce stray AC into the amplifier by means of the speaker cables.

Connectors

A *connector* is a device intended for providing and maintaining good electrical contact between two different sections of a wire or cable. It is important to use the proper connector for a given application. Often, the choice is dictated by convention.

Single-wire connectors are simple. Multiconductor-cable connectors are more complex. Here are several of the most common types of wire and cable connectors used in electrical systems.

CONVENTIONAL PLUG AND OUTLET

We're all familiar with the plugs attached to the ends of cords used with electrical appliances. These have two or three prongs, which fit into outlets with receptacles in the same configuration. The plugs are called *male connectors*, and the receptacles into which the plugs fit into are called *female connectors*. This terminology is universal among connectors.

   Conventional plugs and outlets provide temporary electrical connections. They are not reliable in long-term applications because the prongs on the plug, and the metallic contacts inside the receptacle, tend to *oxidize* or *corrode* over time. A new plug has shiny metal blades, but an old one often has blades that have darkened. Conventional plugs and outlets are more reliable for indoor use than for outdoor use. They are used almost exclusively for utility AC.

CLIP LEADS

A *clip lead* is a short length of flexible wire, equipped at one or both ends with a simple, temporary connector. Clip leads are not suitable for permanent installations, especially outdoors, because they can corrode easily, and the connector can slip out of position. The current-carrying capacity is limited. Clip leads are used primarily in DC and low-frequency AC applications.

   For testing and experimentation when temporary connections are needed, *alligator clips* are often used. They require no modification to the circuit under test. The name derives from the serrated edges of the clip, resembling the mouth of an alligator. Alligator clips come in a wide range of sizes. They can be clamped to an electrical terminal, a metal pipe, or a length of exposed wire.

BANANA CONNECTORS

A *banana connector* is a convenient single-lead connector that slips easily in and out of its receptacle. The banana plug (male connector) actually looks something like a banana, and that is how it gets its name (Fig. 9-7).

   Banana jacks (female receptacles) are often found inside the screw terminals of low-voltage DC power supplies. If frequent changing of leads is necessary, banana connectors make it more convenient than the repeated screwing and unscrewing of the terminals. Banana connectors are electrically reliable, and they have high current-handling capacity. They are not suitable for use at high voltages, because of the danger of electrical shock from the exposed conductors.

HERMAPHRODITIC CONNECTORS

A *hermaphroditic connector* is an electrical plug/jack that has some male contacts and some female contacts. Sometimes both the connectors appear

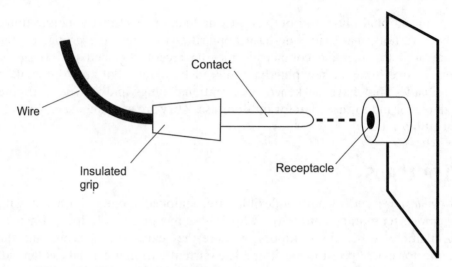

Fig. 9-7. Banana plugs and jacks are convenient for low-voltage DC use.

identical. Since they are equipped with special pins and holes, they can be inserted into each other in only one way. This makes them useful in polarized circuits, such as DC power supplies.

PHONE JACK AND PLUG

A *phone jack and plug* (Fig. 9-8) is a connector pair originally designed for patching telephone circuits, now widely used in electrical and electronic systems.

In its conventional form, the male plug (shown at A) has a rod-shaped metal neck or sleeve that serves as one contact, and a spear-shaped or a ball-shaped metal tip that serves as the other contact. The sleeve and the tip are separated by a ring of insulation. Typical diameters are 1/8 in. (3.175 mm) and 1/4 in. (6.35 mm). The female jack (B) is equipped with contacts that mate securely with the plug when the plug is inserted.

The original phone jack/plug was designed for use with 2-conductor cables. But in recent years, 3-conductor connectors have become common. They are used in high-fidelity stereo sound systems, audio tape recorders, and in the audio circuits of multimedia computers, radio receivers, transmitters, and transceivers. The male plug has a sleeve broken into two parts along with a tip, and the female connector has an extra contact that touches the second sleeve when the plug is inserted.

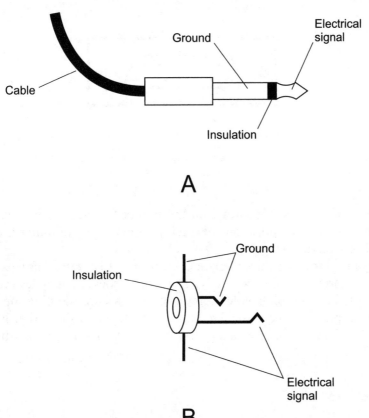

Fig. 9-8.   At A, a 2-conductor phone plug. At B, a 2-conductor phone jack.

PHONO JACK AND PLUG

A *phono jack and plug* is designed for ease of connection and disconnection for use with a coaxial cable. The plug is pushed on and pulled off. It is, in effect, a shielded banana jack/plug. Phono connectors are used in the same applications as phone jacks and plugs.

D-SHELL CONNECTORS

Data cables, of the type used in computer systems, have several (sometimes many) wires. If there are more than three or four wires in a cable, a *D-shell connector* is often used at either end. The D-shell connectors

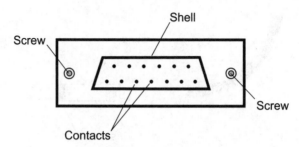

Fig. 9-9. The D-shell connector is recognizable by its shape. The number of contacts can vary.

come in various sizes, depending on the number of wires in the cable. The serial and the parallel ports on the back of a personal computer are almost always D-shell connectors.

The connector has a characteristic appearance (Fig. 9-9). This shape forces the user to insert the plug correctly. The female socket has holes into which the pins of the male plug slide. Screws or clips secure the plug, once it has been put in place. Some D-shell connectors have metal shells that help keep out dust and moisture, and also serve to maintain shielding continuity, if needed.

? PROBLEM 9-5

You have an appliance with a 2-conductor plug on the end of its cord. One of the blades on the plug is slightly wider than the other. The electrical outlets in your house are all of the 3-wire type, with two vertical slots and one round or D-shaped hole below the slots. In each outlet, the left-hand slot is slightly wider than the right-hand slot (when viewed so that the round or D-shaped hole is below the slots). You can plug the 2-wire cord into the 3-wire outlet, but the plug only goes in one way: the wider blade into the wider slot, and the narrower blade into the narrower slot. What is the reason for the difference in the width of the blades on the plug and the slots in the outlets?

✔ SOLUTION 9-5

The 2-wire plug with unequal-width blades is known as a *polarized plug*. The wider blade is meant to be grounded, while the narrower blade is meant to carry the single-phase AC. In a properly wired home electrical circuit, the wider slot in each outlet should be connected to the system ground, and the narrower slot should carry the single-phase AC electricity.

? **PROBLEM 9-6**

You are renting an apartment with hookups for electric laundry machines. You have no laundry machines yet, so you use a laundromat instead. You have noticed that the electrical outlet for the machines looks different than the regular outlets that are found in other places throughout the house. The laundry outlet has three slots. The two on the top are oriented at a slant, and the one on the bottom is vertical. What type of an outlet is this?

✔ **SOLUTION 9-6**

Outlets of this sort normally carry 234 V rms AC. The vertical slot on the bottom is at electrical ground. The slot on the left carries 117 V rms AC with respect to ground. The slot on the right also carries 117 V rms AC with respect to ground, but in the opposite phase (that is, 180° out of phase) from the AC in the slot on the left. The voltage between the two slanted slots is twice the usual 117 V rms AC, or 234 V rms AC.

WHY THEY DON'T CANCEL

In this case, the two AC voltages, although of equal frequency, equal amplitude, and opposite phase, do not cancel, as you might expect at first. They add instead, because the output is taken between them—that is, one with respect to the other—rather than by connecting them together and then taking the output with respect to electrical ground. It's like two cars driving toward each other in exactly opposite directions, each one going 100 km/h. Their average or composite speed is zero, but their speed relative to each other is 200 km/h.

If the two 117 V sources in a 234 V utility outlet were directly connected to each other, the voltages would indeed cancel out. But this would not last long. This action would short out both 117 V circuits, and this would blow the fuses or breakers! It would also create a gigantic spark at the outlet for the brief instant that massive current flowed between the sources. (Don't try this as an experiment. It could cause a fire, and the spark could actually injure you.)

? **PROBLEM 9-7**

If a plug becomes corroded or oxidized, so it no longer makes reliable electrical contact with outlets for which it is designed, what can you do?

✓ **SOLUTION 9-7**

You can usually tell when a plug is corroded or oxidized, because the metal that makes up the prongs is dark or discolored. To remedy this, the prongs can be sanded with fine-grain sandpaper and then wiped off with a dry cloth. Alternatively, an emery board, such as the type sold for filing down broken fingernails, can be used. The prongs should be sanded until the bright metal shows everywhere. If the prongs are too small or too closely spaced for sanding to be practical, then a special contact cleaner can be used. This type of cleaner is available in most good hardware stores or hobby-electronics shops.

Quiz

This is an "open book" quiz. You may refer to the text in this chapter. A good score is 8 correct answers. Answers are in Appendix 1.

1. The insulation in a coaxial cable is also known as the

    (a) shield.
    (b) center conductor.
    (c) braid.
    (d) dielectric.

2. Suppose you have two lengths of AWG No. 16 solid, soft-drawn copper wire. Call them wire X and wire Y. Suppose wire X is twice as long as wire Y. Which of the following statements is true?

    (a) Wires X and Y have the same resistance.
    (b) Wires X and Y have the same resistivity.
    (c) Wire X can safely carry only half the 60-Hz current of wire Y.
    (d) Wire X can safely carry twice the 60-Hz current of wire Y.

3. When numerous wires are bundled into a cable, the maximum safe current-carrying capacity of any particular one of the wires

    (a) is the same as it would be if the wire was in open air all by itself.
    (b) is greater than it would be if the wire was in open air all by itself.
    (c) is less than it would be if the wire was in open air all by itself.
    (d) depends on the length of the cable.

4. A parallel cable

    (a) can generally be longer than other types of cable, because all its conductors are connected in parallel.
    (b) must have all its conductors shorted out at each end, so the wires are all connected in parallel.
    (c) must run parallel to any other nearby cables.
    (d) carries current or signals along multiple, independent paths between two specified points.

5. Suppose you want to connect a hi-fi stereo amplifier to a computer in order to get high-level audio to drive a pair of big speakers. You want to ensure that the amplifier receives the minimum possible amount of external electromagnetic energy from the input cables. The best type of cable to use for each stereo channel (left and right) between the computer and the amplifier is

    (a) a lamp cord.
    (b) a heavy-gauge, insulated, solid copper wire.
    (c) a 3-wire electrical cord.
    (d) a coaxial cable.

6. When two wires of the same gauge must be spliced, a Western Union splice

    (a) has better tensile strength than a twist splice.
    (b) does not provide sufficient electrical continuity.
    (c) works only with stranded wire.
    (d) should never be soldered.

7. When splicing a multiconductor cable, extra electrical isolation for each conductor from the others can be obtained by

    (a) soldering each splice after wrapping it with an electrical tape.
    (b) using twist splices for some conductors and Western Union splices for others.
    (c) splicing each conductor at slightly different points along the cable.
    (d) Any of the above

8. Suppose two wires are both made of a solid, soft-drawn copper. Neither is coated with any sort of insulation or enamel. One is AWG No. 12 and the other is AWG No. 14. Which of the following statements is false?

   (a) The AWG No. 12 wire has a lower DC resistivity than the AWG No. 14 wire.

   (b) The AWG No. 12 wire has a higher tensile strength than the AWG No. 14 wire.

   (c) The AWG No. 12 wire has smaller cross-sectional area than the AWG No. 14 wire.

   (d) The AWG No. 12 wire is a better 60-Hz AC electrical conductor than the AWG No. 14 wire.

9. The center conductor in a coaxial cable

   (a) provides shielding.
   (b) carries the current or signal.
   (c) should always be grounded.
   (d) forms part of the dielectric.

10. As the gauge of a wire increases, the maximum current that it can safely carry

   (a) increases.
   (b) does not change.
   (c) decreases.
   (d) depends on how long it is.

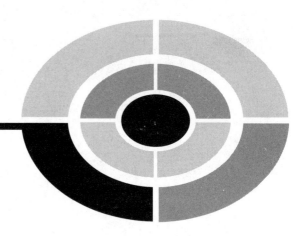

Test: Part Two

Do not refer to the text when taking this test. You may draw diagrams or use a calculator, if necessary. A good score is at least 30 answers (75% or more) correct. Answers are in the back of the book. It's best to have a friend check your score the first time, so you won't memorize the answers, if you want to take the test again.

1. Fill in the blank to make the following sentence true: "As the electrical power demanded from a generator increases, it takes _____ torque (rotational force) to turn the generator shaft."

    (a) reduced
    (b) increased
    (c) reversed
    (d) alternating
    (e) magnetic

2. Suppose a DC component of +30 V is superimposed on 60-Hz, sine-wave, single-phase AC electricity supplied at 330 V pk–pk. What is the peak-to-peak voltage of the resulting electricity?

    (a) 300 V pk–pk
    (b) 330 V pk–pk

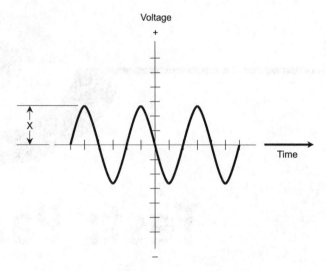

Fig. Test2-1. Illustration for Part Two Test Questions 4 and 5.

(c)   360 V pk–pk
(d)   165 V pk–pk
(e)   117 V pk–pk

3.   The average amplitude of an AC sine wave with no DC component is

(a)   zero.
(b)   half the peak amplitude.
(c)   0.354 times the peak amplitude.
(d)   0.354 times the peak-to-peak amplitude.
(e)   the same as the effective amplitude.

4.   In Fig. Test2-1, the displacement marked X portrays

(a)   the rms voltage.
(b)   the positive peak voltage.
(c)   the negative peak voltage.
(d)   the peak-to-peak voltage.
(e)   the average voltage.

5.   In Fig. Test2-1, suppose each horizontal division represents 1 ms (0.001 s). The frequency of the sine wave is

(a)   4000 kHz.
(b)   400 kHz.
(c)   40 kHz.

(d) 4 kHz.

(e) None of the above

6. The overall diameter of a wire conductor is defined according to a number called the

(a) area.

(b) resistivity.

(c) conductivity.

(d) gauge.

(e) carrying capacity.

7. The point on a sine wave at which the value is at its positive maximum is sometimes called a

(a) trough.

(b) node.

(c) phase point.

(d) zero point.

(e) crest.

8. If an electrolytic capacitor is connected with the polarity wrong, what can happen?

(a) the capacitor might act as a diode.

(b) the capacitor might convert DC to AC.

(c) the capacitor might explode.

(d) the capacitor might convert AC to DC.

(e) Nothing bad. Polarity does not matter with an electrolytic capacitor.

9. Imagine two pieces of wire, each of the same length. Both the wires are AWG No. 14, and both are made of pure copper. One piece of wire is stranded, and the other is solid. Which piece of wire has the greater resistance for DC and 60-Hz utility AC?

(a) The piece of stranded wire.

(b) The piece of solid wire.

(c) Neither; they have the same resistance.

(d) It depends on the actual length of the wires.

(e) More information is needed to answer this question.

10. Suppose an AC source, a load, an AC ammeter, and an AC voltmeter are connected, as shown in Fig. Test2-2. The voltmeter and the ammeter both measure rms values. Assume the voltage source has

no DC component, and is a pure AC sine wave. If the voltmeter indicates 10 V and the ammeter indicates 2 A, what is the power consumed by the load?

(a)   More information is needed to answer this question.
(b)   200 W
(c)   50 W
(d)   20 W
(e)   5 W

11.   Suppose an AC source, a load, an AC ammeter, and an AC voltmeter are connected as shown in Fig. Test2-2. The voltmeter and the ammeter both measure rms values. Assume the voltage source has no DC component, and is a pure AC sine wave. If the voltmeter indicates 117 V and the ammeter indicates 500 mA, what is the resistance of the load?

(a)   58.5 kΩ
(b)   58.5 Ω
(c)   27.4 kΩ
(d)   234 Ω
(e)   468 Ω

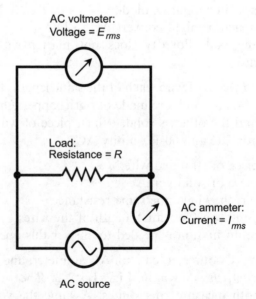

Fig. Test2-2.   Illustration for Part Two Test Questions 10 through 14.

12. Suppose an AC source, a load, an AC ammeter, and an AC volt-meter are connected, as shown in Fig. Test2-2. The voltmeter and the ammeter both measure rms values. Assume the voltage source has no DC component, and is a pure AC sine wave. The load resistance is known to be 117 Ω. What is the rms current through it?

    (a)   1 A rms
    (b)   117 A rms
    (c)   8.55 mA rms
    (d)   0 A rms
    (e)   More information is needed to answer this question.

13. Suppose an AC source, a load, an AC ammeter, and an AC volt-meter are connected, as shown in Fig. Test2-2. The voltmeter and the ammeter both measure rms values. Assume the voltage source has no DC component, and is a pure AC sine wave. The load resistance is known to be 117 Ω. What is the average voltage of the AC source?

    (a)   1 V avg
    (b)   117 V avg
    (c)   8.55 E avg
    (d)   0 V avg
    (e)   More information is needed to answer this question.

14. Suppose an AC source, a load, an AC ammeter, and an AC voltmeter are connected, as shown in Fig. Test2-2. The voltmeter and the ammeter both measure rms values. Assume the voltage source has no DC component, and is a pure AC sine wave. The load resistance is known to be 117 Ω. Then, suddenly, the load burns out, and an open circuit appears in its place. What happens to the voltmeter reading?

    (a)   It drops to 0.
    (b)   It does not change.
    (c)   It increases somewhat.
    (d)   It decreases somewhat.
    (e)   More information is needed to answer this question.

15. Suppose an AC wave has a frequency of 200 kHz. What is the period in microseconds?

    (a)   0.5 μs
    (b)   5 μs

(c)   50 µs

(d)   500 µs

(e)   More information is needed to answer this question.

16. Suppose an AC wave has a peak-to-peak voltage of 200 mV. What is the period in microseconds?

(a)   0.5 µs

(b)   5 µs

(c)   50 µs

(d)   500 µs

(e)   More information is needed to answer this question.

17. What is the advantage of a half-wave rectifier circuit over a full-wave rectifier circuit that uses a transformer with a center-tapped secondary? Assume that the required output voltage and current are the same in either case.

(a)   The half-wave circuit is easier on the transformer.

(b)   The half-wave circuit has better voltage regulation.

(c)   The half-wave circuit costs less.

(d)   The half-wave circuit has an output that is easier to filter.

(e)   Forget it! A half-wave rectifier has no advantages at all.

18. Suppose two rectifier circuits deliver pulsating DC with 60-Hz AC input. One of the rectifier circuits is a full-wave center-tap arrangement. The other is a full-wave bridge arrangement. How do the ripple frequencies of the pulsating DC output from the two circuits compare?

(a)   The ripple frequency at the output of the center-tap circuit is 4 times the ripple frequency at the output of the bridge circuit.

(b)   The ripple frequency at the output of the center-tap circuit is twice the ripple frequency at the output of the bridge circuit.

(c)   The ripple frequency at the output of the center-tap circuit is the same as the ripple frequency at the output of the bridge circuit.

(d)   The ripple frequency at the output of the center-tap circuit is half of the ripple frequency at the output of the bridge circuit.

(e)   The ripple frequency at the output of the center-tap circuit is 1/4 of the ripple frequency at the output of the bridge circuit.

19. Suppose two AC sine waves have the same frequency and the same peak-to-peak voltage, but they are 180° out of phase. The peak-to-peak voltage of one wave with respect to the other is

    (a) equal to 1/4 of the peak-to-peak voltage of either wave.
    (b) equal to half of the peak-to-peak voltage of either wave.
    (c) equal to the peak-to-peak voltage of either wave.
    (d) equal to twice the peak-to-peak voltage of either wave.
    (e) equal to 4 times the peak-to-peak voltage of either wave.

20. Voltage regulation in a DC power supply can be accomplished using

    (a) a Zener diode in parallel with the output.
    (b) a capacitor in parallel with the output.
    (c) a filter choke in parallel with the output.
    (d) a battery in series with the output.
    (e) Any of the above

21. If an AC sine wave has no DC superimposed on it, then the peak-to-peak voltage is equal to

    (a) twice the positive peak voltage with the plus sign eliminated.
    (b) half the positive peak voltage with the plus sign eliminated.
    (c) 0.354 times the positive peak voltage with the plus sign eliminated.
    (d) 2.828 times the positive peak voltage with the plus sign eliminated.
    (e) the positive peak voltage with the plus sign eliminated.

22. A diode failure can occur in a poorly designed power supply as a result of the current surge, when AC is first applied to the input. The risk of such a diode failure can be minimized by

    (a) using diodes with a current rating of many times the normal operating level.
    (b) connecting several diodes in parallel wherever a diode is called for, and using equalizing resistors in series with each diode.
    (c) using a circuit that applies a reduced AC voltage to the transformer primary for 1 or 2 seconds, and then applies the full input voltage.
    (d) Any of the above (a), (b), or (c)
    (e) None of the above (a), (b), or (c)

23. The use of a coaxial cable between the amplifier and the speakers of a hi-fi sound system

   (a) can boost the treble output of the speakers by reducing the resistance in the cable.
   (b) can prevent damage to the speakers by reducing the likelihood of excessive amplifier output.
   (c) can reduce the risk that an electromagnetic field will enter the amplifier by means of the speaker cables.
   (d) can improve the fidelity of the system because coaxial cable is better for DC than regular speaker cable.
   (e) offers no benefit whatsoever and is a waste of money.

24. A D-shell connector would most likely be found

   (a) on the back of a computer, for connecting a cable to an accessory.
   (b) on the end of a lamp cord, for plugging into a utility outlet.
   (c) on the end of a length of coaxial cable, for attachment to a radio receiver.
   (d) on the end of a single-wire cable, for use as a jumper to start a car.
   (e) in the wall of your house, for use as an electrical outlet.

25. The rms voltage of an AC wave with no DC component, but irrespective of the waveform, can never be greater than

   (a) the positive peak voltage.
   (b) the absolute value of the negative peak voltage.
   (c) the peak-to-peak voltage.
   (d) All of the above (a), (b), and (c)
   (e) the average voltage.

26. What is the circuit shown in Fig. Test2-3?

   (a) An unregulated power supply using a half-wave rectifier.
   (b) An unregulated power supply using a full-wave bridge rectifier.
   (c) An unregulated power supply without any filtering circuit.
   (d) A regulated power supply without any rectifier circuit.
   (e) A regulated voltage-doubler power supply.

27. The coil in series with the output of the circuit in Fig. Test2-3, that appears to the right of the parallel-connected capacitor, is

   (a) a transformer.
   (b) an alternator.
   (c) a rectifier.

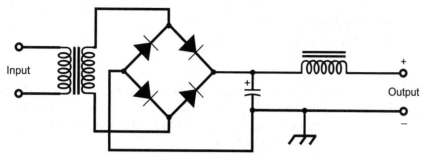

Fig. Test2-3.   Illustration for Part Two Test Questions 26 and 27.

   (d)   a surge suppressor.
   (e)   a filter choke.

28.   An electrolytic capacitor, when used as the filter in a power supply
      having a half-wave rectifier, should

   (a)   have the smallest possible value.
   (b)   be connected in series with the output.
   (c)   be connected with the correct polarity.
   (d)   be connected across the transformer primary.
   (e)   short-circuit the diode.

29.   Figure Test2-4 is a cross-sectional view of a cable. What type of cable is
      this?

   (a)   Lamp cord.
   (b)   Coaxial cable.

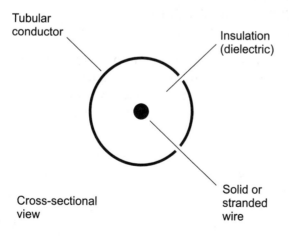

Fig. Test2-4.   Illustration for Part Two Test Question 29.

(c)   Single-wire cable.

(d)   Diametric cable.

(e)   Ribbon cable.

30. Suppose the AC input voltage to a step-up transformer is 180 V rms. The primary-to-secondary turns ratio, T, is 1:3. What is the rms voltage across the entire secondary winding? Assume the input is a sine wave with no DC component.

(a)   0 V rms

(b)   20 V rms

(c)   60 V rms

(d)   540 V rms

(e)   1620 V rms

31. Ribbon cable with many conductors is most likely to be found

(a)   between an electric lamp and the utility outlet.

(b)   in long spans of utility wire running across the countryside.

(c)   inside a personal computer or other high-tech system.

(d)   between a radio transmitter and its antenna.

(e)   Nowhere. There is no such thing as "ribbon cable."

32. A utility plug with darkened blades is probably

(a)   open-circuited.

(b)   short-circuited.

(c)   oxidized.

(d)   fit for use only with DC.

(e)   fit for use at higher voltages than normal.

33. Suppose you have several lengths of solid, soft-drawn, insulated copper wire. The wires are of various gauges in the AWG system. The gauge of each wire is marked clearly on the insulation. Which wire has the highest resistivity for 60-Hz AC, assuming none of them is broken inside the insulation?

(a)   The wire with the highest AWG number.

(b)   The wire with the lowest AWG number.

(c)   It does not make any difference.

(d)   It depends on the relative lengths of the wires.

(e)   There is no way to tell except to measure it with a lab instrument.

34. The most common expression of an effective voltage in an AC utility circuit is

    (a) the positive peak voltage.
    (b) the negative peak voltage.
    (c) the peak-to-peak voltage.
    (d) the root-mean-square voltage.
    (e) the average voltage.

35. An electrical plug with only two blades, one wider than the other, is called a polarized plug. When this type of plug is inserted into a properly wired polarized outlet, the wider blade is connected to

    (a) 117 V rms AC.
    (b) 234 V rms AC.
    (c) DC.
    (d) electrical ground.
    (e) nothing at all; it is "floating."

36. An asset of a twist splice is the fact that

    (a) it is mechanically stronger than other types of splice.
    (b) it can be used with coaxial cable.
    (c) it can carry higher currents than other types of splice.
    (d) it makes a convenient temporary connection between wires.
    (e) it works at higher AC frequencies than other types of splice.

37. If the frequency of an AC sine wave is cut to 1/4 its previous value, and if nothing else about it changes, then the period

    (a) doubles.
    (b) is cut in half.
    (c) quadruples.
    (d) is cut to 1/4 its previous value.
    (e) stays the same.

38. Consider a step-down transformer with a primary-to-secondary turns ratio of 2:1. The input wave is a sine wave with a DC component superimposed. If the input voltage is 160 V rms, then the rms output voltage

    (a) is 0.354 × 160 V rms.
    (b) is 0.707 × 160 V rms.
    (c) depends on the voltage of the DC component at the input.
    (d) is 1.414 × 160 V rms.
    (e) is 2.828 × 160 V rms.

39. If a 3 A slow-blow fuse is used to replace a blown-out 5 A slow-blow fuse in a power supply, the 3 A fuse

    (a)   might not adequately protect the equipment connected to the supply.
    (b)   might blow out needlessly and repeatedly.
    (c)   might short out the input to the power supply.
    (d)   might short out the output of the power supply.
    (e)   might cause degraded voltage regulation.

40. If a 5 A slow-blow fuse is used to replace a blown-out 5 A quick-break fuse in a power supply, the slow-blow fuse

    (a)   might not adequately protect the equipment connected to the supply.
    (b)   might blow out needlessly and repeatedly.
    (c)   might short out the input to the power supply.
    (d)   might short out the output of the power supply.
    (e)   might cause degraded voltage regulation.

PART THREE

Magnetism

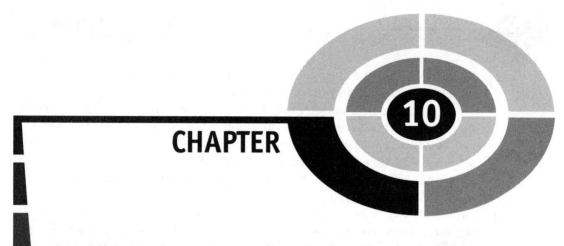

CHAPTER 10

What Is Magnetism?

Electrical and magnetic phenomena are closely related. *Magnetism* exists whenever electric *charge carriers* move relative to other objects, or relative to a frame of reference.

Magnetic Force

As children, most of us discovered that certain metal objects, called *magnets*, "stick" to iron and steel. Iron, nickel, and alloys containing either or both of these elements are known as *ferromagnetic materials*. Magnets are attracted to objects made from these metals or alloys. This attraction does not occur between magnets and metals such as copper or aluminum, unless an electric current flows through the metal. Electrically insulating substances such as wood, plastic, and glass do not attract magnets under normal conditions.

CAUSE AND STRENGTH

When a magnet is brought near a piece of ferromagnetic material, the atoms in the material become lined up, so that the material is temporarily magnetized. This produces a *magnetic force* between the atoms of the ferromagnetic substance and those in the magnet.

If a magnet is near another magnet, the force is stronger than it is when the magnet is near a ferromagnetic substance. The force between two magnets can be either repulsive (the magnets repel, or push away from each other) or attractive (the magnets attract, or pull towards each other), depending on the way the magnets are oriented. Every magnet has poles called *north* (N) and *south* (S), just as a battery has poles called positive (+) and negative (−). As with electrically charged objects, like magnetic poles (N–N or S–S) produce repulsive force, and opposite poles (S–N or N–S) produce attractive force. The force, whether repulsive or attractive, grows stronger as the magnets are brought closer together.

Some magnets are so strong that no human being can pull them apart, if they get stuck together with poles S–N or N–S, and no person can push a pair of them all the way together against their mutual repulsive force N–N or S–S. Other magnets are so weak that the forces cannot be noticed, except with sensitive lab instruments.

CHARGE CARRIERS IN MOTION

Whenever the atoms in a piece of ferromagnetic material are aligned, a *magnetic field* exists around that piece of material. A magnetic field can also be produced by the motion of electric charge carriers, either in a wire or in free space. The charge carriers are usually electrons, but moving protons, atomic nuclei, or charged atoms can also produce magnetic fields.

The magnetic field around a *permanent magnet* arises from the same cause as the field around a wire that carries an electric current. The responsible factor in either case is the motion of electrically charged particles. In a wire, the electrons move along the conductor, being passed among the atomic nuclei in one direction or the other. In a permanent magnet, the movement of electrons around the nuclei occurs in such a manner that an *effective current* is produced by their motion within the individual atoms.

LINES OF FLUX

Physicists consider magnetic fields to be made up of *flux lines*, or *lines of flux*. The intensity of the field is determined according to the number of flux lines passing through a flat region having a certain cross sectional area, usually 1 square centimeter (1 cm^2) or 1 square meter (1 m^2). The lines are imaginary, not solid threads or fibers. But they are a useful concept in defining the geometry of a magnetic field.

Have you seen a demonstration in which iron filings are placed on a sheet of paper, and then a magnet is placed underneath the paper? The filings arrange themselves in a pattern that shows the shape of the magnetic field in the vicinity of the magnet. A bar magnet has a field whose lines of flux have a characteristic pattern, shown in Fig. 10-1. In this illustration, the flux lines are the dashed curves. The flux lines converge or diverge at the ends of the magnet, where the poles are centered.

Another experiment involves passing a current-carrying wire through a horizontally oriented piece of paper so the wire is perpendicular to the paper, and then scattering iron filings on the paper. The filings become grouped in circles. This shows that the lines of flux are circular as viewed through any plane passing through the wire at a right angle. The flux circles are centered

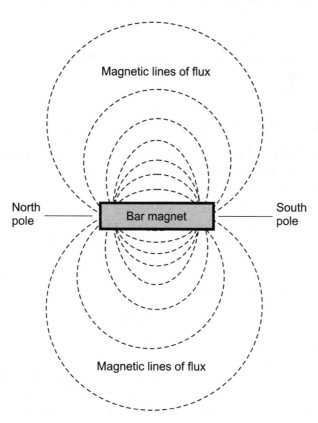

Fig. 10-1. Magnetic flux (dashed curves) around a bar magnet (rectangle).

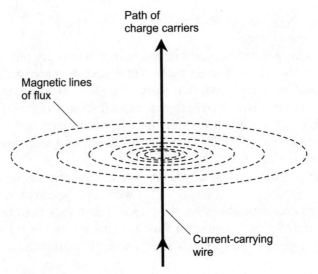

Fig. 10-2. Magnetic flux (dashed curves) produced by charge carriers traveling in a straight line.

on the axis of the wire, or the path along which the charge carriers move, as shown in Fig. 10-2.

POLARITY

A magnetic field has a direction, or an orientation, at any point in space near a current-carrying wire or a permanent magnet. The flux lines run parallel to the direction of the field. A magnetic field is considered to begin, or originate, at a north pole, and to end, or terminate, at a south pole. Usually, magnetic lines of flux tend to diverge from the north poles and converge toward the south poles. But there are certain exceptions, such as in the case of flux surrounding a straight wire, as we'll see in a moment.

The north and the south poles of a permanent magnet are not the same as the north and the south magnetic poles of the earth (the *geomagnetic poles*). They are the opposite! The north geomagnetic pole is in reality a south magnetic pole, because it attracts the north poles of permanent magnets. The south geomagnetic pole is actually a north magnetic pole, because it attracts the south poles of permanent magnets.

With a bar magnet, it is apparent where the magnetic poles are located. In the case of a current-carrying wire the locations of the poles are not so obvious, because the magnetic field goes around and around like a dog chasing its own tail.

AMPERE'S LAW

Remember, physicists define theoretical current, also called conventional current, as flowing from the positive electric pole to the negative electric pole (plus to minus). This is opposite from the direction that electrons move. Suppose the conventional current portrayed in Fig. 10-3 flows out of the page toward you. This means that the region above the page is electrically negative relative to the region below the page. According to a rule called *Ampere's Law*, the direction of the magnetic flux is counterclockwise in this situation. When conventional current flows toward you, the resulting magnetic flux turns counterclockwise from your point of view.

Ampere's Law is sometimes called the *right-hand rule*. If you hold your right hand with the thumb pointing out straight and the fingers curled, and then point your thumb in the direction of the conventional current flow in a straight wire, your fingers curl in the direction of the magnetic flux. Similarly, if you orient your hand so that your fingers curl in the direction of the magnetic flux and then straighten out your thumb, your thumb points in the direction of the conventional current.

If you want to determine the direction of magnetic flux flow relative to the flow of electrons (that is, from minus to plus), use your left hand instead of your right hand.

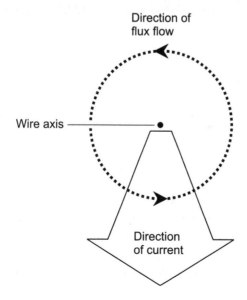

Fig. 10-3. Ampere's Law defines the direction in which magnetic flux "flows."

MONOPOLES AND DIPOLES

A charged electric particle, hovering in space and not moving, is an *electric monopole*. This means it has only one pole. The *electric flux lines* around an electric monopole aren't closed. A positive charge does not have to be mated with a negative charge, although it can be, and often is. When there is a positive electric pole near a negative pole, an *electric dipole* exists.

It is tempting to think that the magnetic field around a current-carrying wire is caused by a monopole, or that there aren't any poles at all, because the concentric circles apparently don't originate or terminate anywhere. They don't converge or diverge. However, there is a way to define the poles. Imagine two distinct points on one of the circular magnetic lines of flux in a perpendicular plane intersecting the wire. A *magnetic dipole*, or a pair of opposite magnetic poles, is formed by the flux going from one point to the other. This can be called a "virtual magnet" because it isn't a physical piece of metal, but only a geometric line segment connecting two points in space (Fig. 10-4). The flux moves out from the north pole and goes toward the south pole. In this drawing, the conventional current is from left to right; the electrons move from right to left.

The lines of flux in the vicinity of a magnetic dipole always connect the two poles. Some flux lines are straight in a local sense, but in the larger sense, they are always curves. The greatest magnetic field strength around a permanent magnet is near the poles, where the flux lines converge or diverge. Around a current-carrying wire, the greatest magnetic field strength is near the wire.

? **PROBLEM 10-1**

Suppose a way is found to create a magnetic monopole. Imagine a point that is a magnetic pole, but there is no opposite pole anywhere nearby. What would the lines of flux around such a point look like?

✓ **SOLUTION 10-1**

The lines of flux would be straight, and they would radiate outward from (or converge inward towards) the point in 3 dimensions. Figure 10-5 is a 2-dimensional cross sectional diagram of this situation. The magnetic flux is shown as dashed lines. If a magnetic monopole were placed on the page, the page were oriented horizontally, and iron filings were scattered on the page, the filings would orient themselves more or less along the dashed lines.

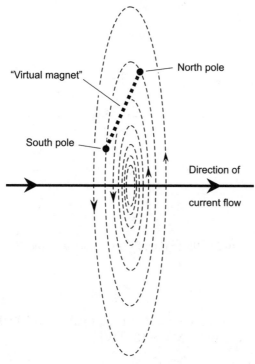

Fig. 10-4. A "virtual magnet" near a current-carrying wire.

[?] **PROBLEM 10-2**

Suppose the magnetic monopole shown in Fig. 10-5 were a north pole. Would the lines radiate outward or converge inward? What if the monopole were a south pole?

[✔] **SOLUTION 10-2**

If the monopole were a north pole, the lines would radiate out from it. If the monopole were a south pole, the lines would converge in towards it.

Magnetic Field Strength

The overall magnitude of a magnetic field is measured in units called *webers*, symbolized Wb. A smaller unit, the *maxwell* (Mx), is sometimes used if a

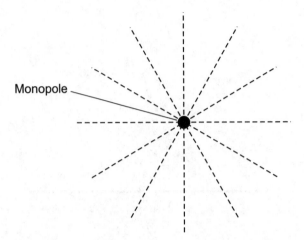

Fig. 10-5. Illustration for Problems 10-1 and 10-2.

magnetic field is very weak. One weber is equivalent to 100,000,000 maxwells. Thus, $1\,\text{Wb} = 10^8\,\text{Mx}$, and $1\,\text{Mx} = 10^{-8}\,\text{Wb}$.

THE TESLA AND THE GAUSS

You will sometimes hear or read about the "strength" of a magnet in terms of webers or maxwells. But more often, you'll hear or read about units called *teslas* (T) or *gauss* (G). These units are expressions of the concentration, or intensity, of the magnetic field within a certain cross section.

The *flux density*, or the number of "flux lines per unit cross-sectional area," is a more useful expression for magnetic effects than the overall quantity of magnetism. Flux density is customarily denoted B in equations. A flux density of one tesla (1 T) is equal to one weber per meter squared ($1\,\text{Wb/m}^2$). A flux density of one gauss (1 G) is equal to one maxwell per centimeter squared ($1\,\text{Mx/cm}^2$). The gauss is equivalent to 0.0001 tesla. That is, $1\,\text{G} = 10^{-4}\,\text{T}$, and $1\,\text{T} = 10^4\,\text{G}$. If you want to convert an expression of magnetic flux density from teslas to gauss (not gausses!), multiply by 10^4. If you want to convert an expression of magnetic flux density from gauss to teslas, multiply by 10^{-4}.

If you are confused by the distinctions between webers and teslas, or between maxwells and gauss, think of a light bulb. Suppose a lamp emits 20 W of visible-light power. If you enclose the bulb completely in a chamber, then 20 W of visible light strike the interior walls of the chamber, no matter how large or small the chamber happens to be. But this is not a very useful notion of the brightness of the light. You know that a 20 W bulb gives plenty of light for a small closet, but it is nowhere near adequate to illuminate a gymnasium. The important consideration is the number of watts *per unit area*. When we say the bulb gives off a certain number of watts of visible light, it's like saying a magnet has an overall flux field of so-many webers or maxwells. When we say that the bulb produces a certain number of watts per unit area, it's analogous to saying that a magnetic field has a flux density of so-many teslas or gauss.

THE AMPERE-TURN AND THE GILBERT

In some situations, units other than the weber, maxwell, tesla, or gauss are preferred. One of these units is known as the *ampere-turn* (At). This is a unit of *magnetomotive force*. A wire, bent into a circle and carrying 1 A of current, produces 1 At of magnetomotive force. If the wire is bent into a loop having 50 turns, and the current stays the same, the resulting magnetomotive force becomes 50 times as great, that is, 50 At. If the current in the 50-turn loop is reduced to 1/50 A or 20 mA, the magnetomotive force goes back down to 1 At.

A unit called the *gilbert* is also used to express magnetomotive force. This unit is equal to about 1.256 At. To approximate ampere-turns when the number of gilberts is known, multiply by 1.256. To approximate gilberts when the number of ampere-turns is known, multiply by 0.796.

These units are most often associated with *electromagnets*, which derive their magnetic "power" from electrical currents. Electromagnets are widely used in industry, in medical devices, and in scientific research and development. We'll see how they work in Chapter 12.

FLUX DENSITY VERSUS CURRENT

In a straight wire carrying a steady, direct current, surrounded by free space (air or a vacuum), the flux density is greatest near the wire, and diminishes with increasing distance from the wire. There is a formula that expresses flux density as a function of distance from the wire. Like all formulas in physics, it is perfectly accurate only under idealized circumstances.

Consider an idealized wire that is infinitely thin and absolutely straight. Suppose it carries a current of I amperes. Consider a point P at a distance r (in meters) from the wire, as measured along the shortest possible route (that is, within a plane perpendicular to the wire). This is illustrated in

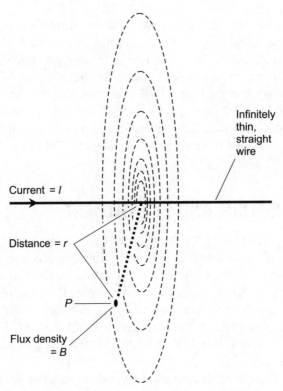

Fig. 10-6.   The flux density B at a point P near a current-carrying wire varies with the current I in the wire, and also with the distance r from the wire.

Fig. 10-6. Let B be the flux density (in teslas) at point P. The following formula applies:

$$B = 2 \times 10^{-7}(I/r)$$

As long as the thickness of the wire is small compared with the distance r from it, and as long as the wire is reasonably straight in the vicinity of the point P at which the flux density is measured, this formula is a good indicator of what happens around real-world electrical conductors carrying current.

? PROBLEM 10-3

What is the flux density in teslas at a distance of 20 cm from a straight, thin wire carrying 400 mA of direct current?

✔ **SOLUTION 10-3**

First, convert the distance to meters and current to amperes. Thus $r = 0.2\,\text{m}$, and $I = 0.4\,\text{A}$. Then plug these numbers into the formula:

$$B = 2 \times 10^{-7}(I/r)$$
$$= 2 \times 10^{-7}(0.4/0.2)$$
$$= 4 \times 10^{-7}\ \text{T}$$

? **PROBLEM 10-4**

In the above scenario, what is the flux density B_{gauss} (in gauss) at point P?

✔ **SOLUTION 10-4**

To figure this out, we must convert from teslas to gauss. That means we must multiply the answer from the previous problem by 10^4:

$$B_{\text{gauss}} = 4 \times 10^{-7} \times 10^4$$
$$= 4 \times 10^{-3}\ \text{G}$$
$$= 0.004\ \text{G}$$

Geomagnetism

The earth has a core made up largely of iron, heated to the extent that some of it is liquid. As the earth rotates, the iron flows in complex ways. This flow gives rise to the natural magnetic field that surrounds our planet and extends millions of kilometers into space.

EARTH'S MAGNETIC POLES AND AXIS

The geomagnetic field has poles, just as a bar magnet does. They are not at the geographic poles, which are the points on the surface through which the earth's *rotational axis* cuts. The *north geomagnetic pole* (which is actually a south magnetic pole, as discussed earlier) is located in extreme northern Canada. The *south geomagnetic pole* is in the ocean near the coast of Antarctica. The *geomagnetic axis* is therefore tilted relative to the axis on which the earth rotates. Besides this, the geomagnetic axis does not run exactly through the center of the earth.

SOLAR WIND

Charged subatomic particles from the sun constantly stream outward through the solar system. This is called the *solar wind*. It distorts the geomagnetic field, in effect "blowing" the magnetic lines of flux out of symmetry. On the side of the earth facing the sun, the geomagnetic lines of flux are compressed. On the side of the earth opposite the sun, the geomagnetic lines of flux are dilated. This distortion is shown in Fig. 10-7. The same effect occurs with the magnetic fields around other planets in the solar system, particularly Jupiter, which has a tremendous planetary magnetic field.

At and near the earth's surface, the lines of flux are nearly symmetrical with respect to the geomagnetic poles. As the distance from the earth increases, the extent of the distortion increases. As the earth rotates, the geomagnetic field does a complicated twist-and-turn dance into space in the direction facing away from the sun.

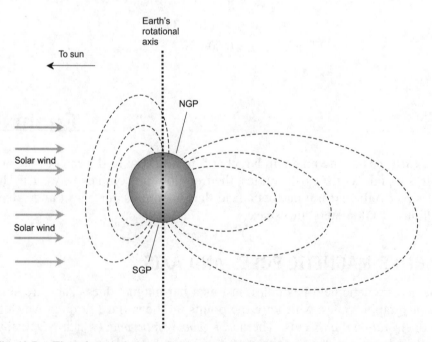

Fig. 10-7. The solar wind "blows" the geomagnetic field into a non-symmetrical shape. Dashed curves depict the lines of flux. The north geomagnetic pole is abbreviated NGP. The south geomagnetic pole is abbreviated SGP.

THE MAGNETIC COMPASS

The earth's magnetic field was noticed even in ancient times. Certain rocks, called *lodestones*, when hung by strings, always orient themselves in a generally north–south direction. Long ago, this was correctly attributed to the presence of a "force" in the atmosphere. It must have seemed like magic. The effect was manifested all over the civilized world, from Alexandria to the Arctic Circle, from the Strait of Gibraltar to the Black Sea, without any visible evidence of its presence!

Long before the reasons for this phenomenon were known, the effect was put to use by seafarers and land explorers. Today, a *magnetic compass* can still be a valuable navigation aid, used by mariners, backpackers, and others who travel far away from familiar landmarks. It can work when more sophisticated navigational devices fail.

The geomagnetic field and the magnetic field around a compass needle interact, so a force is exerted on a little permanent magnet inside the compass. This force works not only in a horizontal plane (parallel to the earth's surface), but also vertically, in most locations. The vertical component of the force is zero at the *geomagnetic equator*, a line running around the globe equidistant from both geomagnetic poles. As the *geomagnetic latitude* increases, either towards the north or the south geomagnetic pole, the magnetic force pulls up and down on the compass needle more and more. The extent of this vertical component at any particular location is called the *inclination* of the geomagnetic field at that location. One end of the needle seems to insist on dipping down toward the compass face, while the other end tilts up toward the glass.

A magnetic compass rarely points exactly toward geographic north, because the north geomagnetic pole doesn't coincide with the north geographic pole. The difference, in degrees, between the horizontal-plane (or *azimuth*) orientation of a compass needle and true geographic north is called the *declination*. The declination varies, depending on where you are. It is $0°$ at locations on a band around the earth passing through both the geomagnetic north pole and the geographic north pole (except for points between the two poles, where the declination is $180°$!). If you try to use a compass at, or near, a geomagnetic pole, the needle will not orient itself in any particular direction, and the compass will be useless. This was not noticed until compass-carrying explorers ventured into the frigid island region of northern Canada.

CHARGE CARRIERS IN SPACE

The particles composing the solar wind carry positive electric charge. Because of their electric charge and their motion, these particles each produce a small effective current. The combined effect of all the moving particles is the equivalent of a huge conventional current, flowing along more or less straight lines emanating outward from the sun. The current produced by each moving, charged particle generates a magnetic field. When the magnetic fields produced by the particles interact with the geomagnetic field, the particles are accelerated toward the geomagnetic poles. They not only move, but they move with varying speed and direction. This gives rise to an *electromagnetic (EM) field*. We'll learn about EM fields in the next chapter.

If there is an eruption on the sun called a *solar flare*, the sun ejects more charged particles than normal. When these particles approach the geomagnetic poles, the resulting magnetic field can disrupt the geomagnetic field. This produces an event called a *geomagnetic storm* that can affect long-distance radio communication and broadcasting. If the fluctuations in the geomagnetic field are intense enough, even wire communications and electric power transmission can be disrupted. The *aurora borealis* (northern lights) and *aurora australis* (southern lights) are frequently observed at night during geomagnetic storms.

[?] **PROBLEM 10-5**

Draw a diagram showing how the geomagnetic axis and the geomagnetic lines of flux would appear, relative to the earth's rotational axis, if there were no solar wind.

[✓] **SOLUTION 10-5**

See Fig. 10-8. Note that the earth has, in effect, a huge bar magnet running through it at a slant, and slightly off center. In this drawing, the north geomagnetic pole (NGP) is actually a south magnetic pole, and the south geomagnetic pole (SGP) is actually a north magnetic pole. The locations of these poles, for practical purposes, can be considered to exist on the earth's surface, the NGP in northern Canada and the SGP near the coast of Antarctica.

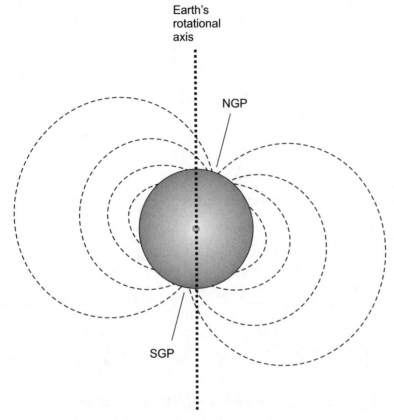

Fig. 10-8.   Illustration for Problem 10-5.

Quiz

This is an "open book" quiz. You may refer to the text in this chapter. A good score is 8 correct answers. Answers are in Appendix 1.

1. In Fig. 10-9, the dashed curves represent

    (a)   electric current.
    (b)   magnetic poles.
    (c)   lines of flux.
    (d)   lines of inclination.

2. In the scenario portrayed by Fig. 10-9, how does B_P compare with B_Q?

    (a) B_P is less than B_Q.
    (b) B_P is the same as B_Q.

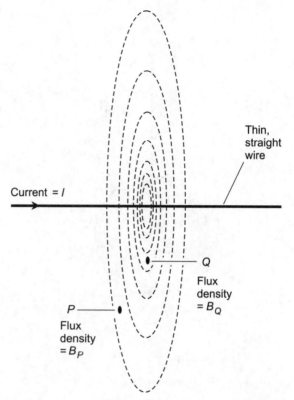

Current = I

Thin, straight wire

Q

Flux density = B_Q

P

Flux density = B_P

Fig. 10-9.   Illustration for Quiz Questions 1 through 4.

(c) B_P is greater than B_Q.

(d)   There is no way to tell without more information.

3.   In the scenario portrayed by Fig. 10-9, suppose you are off to the right
of the diagram, looking down the wire. How does the magnetic flux
appear to flow from this point of view? Assume the arrow on the
wire represents the direction of the conventional current.

(a)   The flux seems to flow clockwise.

(b)   The flux seems to flow counterclockwise.

(c)   The flux seems to flow toward you.

(d)   The flux seems to flow away from you.

4.   In the scenario of Fig. 10-9, suppose the wire carries 4.5 A of current,
and point Q is 0.18 cm away from the wire axis. What is B_Q in gauss
(G)?

(a)   5 G

(b)   162 G

(c)   600 G

(d)   There is no way to tell without more information.

5.   The lines of flux in the vicinity of a bar magnet

(a)   are always straight.

(b)   are always circles.

(c)   always diverge from the center of the magnet.

(d)   always connect the two poles.

6.   Magnetic fields can be caused by all of the following except

(a)   the alignment of the atoms in a ferromagnetic substance.

(b)   a straight conductor carrying an electric current.

(c)   a stationary, charged particle.

(d)   a moving, charged particle.

7.   A stream of charged particles such as protons, hurtling through space at high speed toward the earth, produces

(a)   an effective current.

(b)   a magnetic field.

(c)   effects on the earth's magnetic field.

(d)   All of the above

8.   The direction in which magnetic flux flows near a straight wire, relative to the current that produces it, can be figured out if we know

(a)   the voltage applied to the wire.

(b)   Ohm's Law.

(c)   Ampere's Law.

(d)   the current in amperes and the length of the wire.

9.   The magnetomotive force around a current-carrying, air-core coil can be figured out, if we know

(a)   the voltage applied to the coil.

(b)   Ohm's Law.

(c)   Ampere's Law.

(d)   the current in amperes and the number of turns in the coil.

10.   Which of the following scenarios does not produce a magnetic force of attraction?

(a) The north pole of a permanent magnet brought near a piece of iron.

(b) The north pole of a permanent magnet brought near a piece of glass.

(c) The north pole of a permanent magnet brought near the south pole of another permanent magnet.

(d) All of the above produce scenarios produce a magnetic force of attraction.

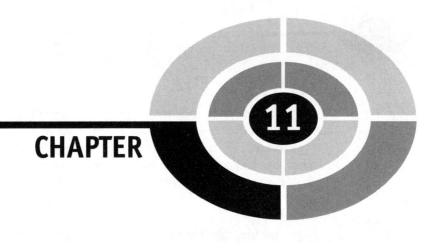

Electromagnetic Effects

Electricity and magnetism are closely related. An electric current can produce a magnetic field, and a magnetic field can induce electric currents in moving conductors. Under some circumstances, magnetic fields can generate electric fields, and vice versa. In this chapter, we'll look at some of the effects common to both electricity and magnetism.

Electromagnetic Deflection

When a magnetic compass is placed near a wire carrying DC, the needle is deflected. The extent of the deflection depends on how close the compass is to the wire, and on how much current the wire carries. This phenomenon is called *galvanism* or the *electromagnetic effect*. When it was first observed, experimenters tested various arrangements to obtain the greatest possible current-detecting sensitivity. When the wire was wrapped in a coil around the compass, with the axis of the coil aligned east–west (and the plane of the coil therefore aligned north–south), the result was a device that could indicate the presence of small currents.

THE GALVANOMETER

A *galvanometer* is a sensitive device for detecting the presence of, and measuring, DC in a wire. It takes advantage of the electromagnetic effect. The galvanometer is similar to an ammeter, but the needle of the galvanometer rests in the center position instead of the left-hand end of a graduated scale. Current flowing in one direction results in deflection of the needle to the right; current in the other direction causes the needle to move to the left.

You can build a galvanometer using a compass and some insulated wire. The best wire for this purpose is called *bell wire*. It is a solid, soft-drawn, insulated copper wire of AWG No. 16 or No. 18, and can be found in any good hardware store. Simply wind the wire several times around the compass along the north–south line on its scale, as shown in Fig. 11-1. Set the compass down on a non-ferromagnetic horizontal surface such as a wooden table, so the arrow floats freely. Turn the compass so the wire coil and the arrow line up with each other. Then connect a small dry cell or a "transistor" battery to the coil for 1 or 2 seconds, and watch the needle deflect. Reverse the polarity of the battery, and try the experiment again; the needle will deflect in the opposite direction. Don't leave the battery connected to the coil for more than a couple of seconds at a time.

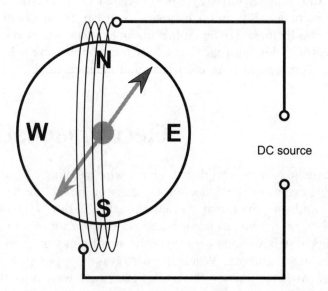

Fig. 11-1. A simple galvanometer can be made by wrapping insulated wire several times around a magnetic compass.

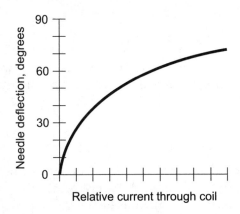

Fig. 11-2.   In a galvanometer, the relationship between coil current and needle deflection is a curve when graphed. The deflection approaches 90° as the current rises without limit.

The compass needle gets deflected because the magnetic field produced by the current in the coil is oriented at a right angle to the earth's magnetic field. When the coil carries DC, the compass "sees" the magnetic poles as combinations of the geomagnetic poles and the poles of the field set up by the coil. The stronger the current in the wire becomes, the greater is the coil's field with respect to the earth's field, and the more the compass gets influenced. For a given number of coil turns, a given compass, and a given physical location, the extent of the needle deflection (in degrees of arc) from magnetic north is proportional to the current, in amperes, carried by the coil. The function is not linear; that is, the relationship between the coil current and the needle deflection is not a straight line. It resembles the graph in Fig. 11-2. Nevertheless, if a milliammeter is available for calibration purposes so the exact deflection-vs-current function can be determined, the galvanometer can serve as a sensitive current meter.

? PROBLEM 11-1

Suppose a galvanometer coil carries 10 mA of DC, causing the needle to be deflected 20° to the west of geomagnetic north. What will the needle do if the direction in which the current flows is reversed, but the current remains at 10 mA?

✔ SOLUTION 11-1

The compass needle will be deflected 20° to the east of geomagnetic north.

? PROBLEM 11-2

What will a galvanometer do if a source of 60 Hz household AC is connected to its coil, with a large-value resistor in series to ensure that the current doesn't heat up the wire too much, or produce a magnetic field that is too strong?

✔ SOLUTION 11-2

The compass needle will point toward the geomagnetic north, just as if there were no current through the coil. A barely perceptible vibration might be observed in the needle, as it attempts to move first to the west of north, then to the east, then to the west again as the magnetic force on it changes direction 120 times per second!

THE ELECTROMAGNETIC CRT

An *electromagnetic cathode-ray tube* (or *electromagnetic CRT*) provides another example of electromagnetic deflection. In the electromagnetic CRT, a "gun" emits a high-intensity stream of electrons. This beam is focused and accelerated as it passes through donut-shaped electrodes called *anodes*, which are positively charged. The motion of the electrons produces an electric current. This current gives rise to a magnetic field, just as the flow of current through a wire produces a magnetic field around the wire. After they have been accelerated by the anodes, the electrons pass through magnetic fields produced by pairs of *deflecting coils* situated at right angles to each other. The electron beam is deflected as it passes the coils, because of the interaction between the magnetic fields from the coils and the magnetic field produced by the moving electrons. The electrons travel on, their direction changed more or less after passing by the coils, until they strike a screen whose inner surface is coated with phosphor. The impact of the electrons causes the phosphor to glow.

Unless current through the deflecting coils moves the electron beam back and forth, up and down, or in some combination of these directions, you'll only see a fixed, bright spot in the center of the CRT screen. Signals applied to the deflecting coils make meaningful displays possible. Figure 11-3 is a simplified functional drawing of an electromagnetic CRT. To minimize clutter, only one pair of *deflecting coils* is shown. As the electrical current through the coils varies in intensity, the magnetic field between them fluctuates in strength, and the electron beam is deflected upward or downward to a greater or lesser extent. If AC is supplied to the coils, then

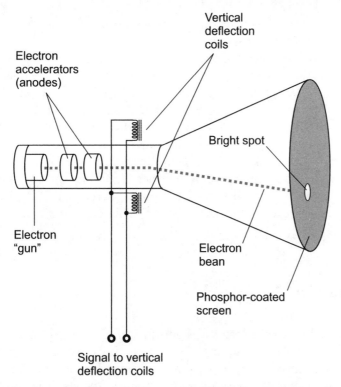

Fig. 11-3.   An electromagnetic cathode-ray tube (CRT). Current in the coils produces a
magnetic field that deflects the electron beam.

the polarity of the magnetic field alternates, and the electron beam is
deflected alternately in opposing directions.

THE CATHODE-RAY OSCILLOSCOPE

The electromagnetic CRT is the heart of an old-fashioned (but still common)
lab instrument called a *cathode-ray oscilloscope*. In this device, the *horizontal
deflecting coils* receive an alternating or fluctuating direct current in the form
of a sawtooth wave, which causes the beam to sweep across the screen at a
controllable, constant speed from left to right, return almost instantaneously,
and then sweep across the screen from left to right again, over and over.
The vertical coils receive a signal to be analyzed. This causes the beam to
be deflected up and down as it passes along from right to left. The waveform
graphs in Chapter 6, such as the sine wave, the square wave, and the

sawtooth wave, were derived from the appearances of these types of currents on the screen of an oscilloscope.

Complex waves can be analyzed using an oscilloscope. The speed at which the beam moves horizontally across the screen can be adjusted, so the instrument can be used to examine waveforms over a wide range of frequencies. The sensitivity of the instrument can be adjusted by switching various resistors into the circuit, in series with the vertical deflection coils. This makes it possible to get almost any waveform, regardless of its intensity or frequency, to fit neatly on the screen. Specialized circuits synchronize the *sweep frequency* at some whole-number fraction of the input-signal frequency, so the waveform stays fixed on the screen instead of constantly racing or jerking toward the left or right. This synchronization process is called *triggering*.

Electromagnetic Fields

Charged particles, such as electrons and protons, are surrounded by *electric fields*. Magnetic poles or moving charged particles produce *magnetic fields*. When the fields are strong enough, they extend a considerable distance. When electric or magnetic fields vary in intensity, the result is an *electromagnetic (EM) field*.

STATIC FIELDS

You have observed the attraction between opposite poles of magnets, and the repulsion between like poles. Similar effects occur with electrically charged objects. These electrical forces seem to operate only over short distances under laboratory conditions, but this is because such fields rapidly weaken, as the distance between poles increases, to less than the smallest intensity we can detect. In theory, an electric field extends into space indefinitely, unless something blocks it, such as a grounded enclosure made of conducting wire mesh or conducting sheet metal.

A permanent magnet, or a wire carrying a constant electric current, produces a magnetic field that rapidly weakens as you get farther and farther from its origin. In theory, a magnetic field extends into space forever, unless something blocks it, such as an enclosure made of ferromagnetic sheet metal.

The existence of a constant voltage difference between two nearby objects, or a constant current in a wire, produces a static electric field (also called an *electrostatic field*) or a static magnetic field (which can be called a *magnetostatic field*, but that term is rarely used). Charged objects and constant currents, all by themselves, don't produce EM fields. For such a thing to happen, the voltage or current must not only be present, but must vary in intensity.

FLUCTUATING FIELDS

A fluctuating current in a wire, or a variable *charge gradient* between two nearby objects, gives rise to an EM field, which is a sort of "dance" between a fluctuating magnetic field and a fluctuating electric field. The varying magnetic field gives rise to a varying electric field, which in turn gives rise to another varying magnetic field. This process keeps on repeating, so the electric and magnetic fields "leapfrog" through space at the speed of light. In a vacuum, this speed, denoted by the lowercase italic letter c, is a well-known constant, equal to approximately 299,792 km/s (186,282 mi/s). This is often rounded off to 3.00×10^8 m/s for use in calculations.

An EM field can travel, or *propagate*, tremendous distances. It dies off more gradually than a static electric field or a static magnetic field. The electric and magnetic lines of flux in an EM field are perpendicular to each other at every point in space. The *direction of propagation* of an EM field is perpendicular to the electric lines of flux, and is also perpendicular to the magnetic lines of flux (Fig. 11-4).

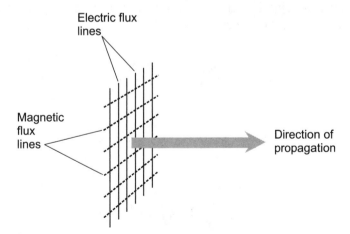

Fig. 11-4.   Propagation of an EM field. At every point in space, the direction in which the EM field travels is perpendicular to both the electric lines of flux and the magnetic lines of flux.

For an EM field to exist, the electric charge carriers must not only be set in motion, but they must also be accelerated. That is, their velocity must be made to change. The most common method of creating this sort of situation is the introduction of an alternating current (AC) in an electrical conductor. It can also result from the deflection of charged-particle beams by electric or magnetic fields.

All wires that carry AC, even household utility wires, produce EM fields. When charged particles from the sun are forced into curved paths near the geomagnetic poles, EM fields are produced. These fields occur over a wide range of frequencies, and can be extremely strong. They are notorious for their tendency to disrupt long-range radio broadcasting and communications.

FREQUENCY AND WAVELENGTH

As the frequency of an EM field in free space increases, the *wavelength* (physical distance between wave crests) decreases, as shown in Fig. 11-5. The frequency and the wavelength are inversely proportional. Their product is a constant that is equal to their speed, which is the speed of light. Figure 11-5A shows a hypothetical EM wave. At B, the frequency is doubled, so the waves are half as long as they are at A. At C, the frequency is quadrupled, and the waves are therefore only 1/4 as long as they are at A.

At 1 kHz, the wavelength of an EM field in free space is about 300 km. At 1 MHz, the wavelength is about 300 m. At 1 GHz, the wavelength is about 300 mm. At 1000 GHz, an EM signal has a wavelength of 0.3 mm, a distance so small that a magnifying glass would be necessary to resolve it if it were something that could be directly seen.

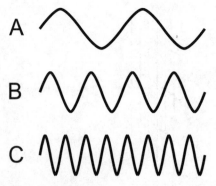

Fig. 11-5. At A, an EM wave in space. At B, the wavelength is cut in half when the frequency is doubled. At C, the wavelength becomes 1/4 as great when the frequency is quadrupled.

The frequency of an EM wave can get much higher than 1000 GHz, and the corresponding wavelength can get much shorter than 0.3 mm. Some known EM waves have wavelengths of only 0.00001 *Ångström* (10^{-5} Å). The Ångström is equivalent to 10^{-10} m, and is used by some scientists to denote extremely short EM wavelengths. A microscope of high magnifying power would be needed to see an object with a length of 1 Å. Another unit, increasingly preferred by scientists these days, is the *nanometer* (nm), where 1 nm = 10^{-9} m = 10 Å.

Wavelength is denoted by the lowercase, italicized Greek letter lambda (λ). Frequency is denoted by the lowercase, italicized English letter f. The wavelength λ, in meters, as a function of the frequency f, in hertz, for an EM field in free space is given by the following formula:

$$\lambda = c/f$$
$$\lambda = 3.00 \times 10^{8}/f$$

This same formula can be used for λ in millimeters and f in kilohertz, for λ in micrometers and f in megahertz, and for λ in nanometers and f in gigahertz. Remember your prefix multipliers: a millimeter (1 mm) is 10^{-3} m, a micrometer (1 μm) is 10^{-6} m, and a nanometer (1 nm) is 10^{-9} m.

The formula for frequency f, in hertz, as a function of the wavelength λ, in meters, for an EM field in free space is:

$$f = c/\lambda$$
$$f = 3.00 \times 10^{8}/\lambda$$

As in the preceding case, this formula will work for f in kilohertz and λ in millimeters, for f in megahertz and λ in micrometers, and for f in gigahertz and λ in nanometers.

MANY FORMS

The discovery of EM fields led to all the wireless communications systems we know today. *Radio waves* were the earliest useful manifestations of the EM radiation, but these waves are by no means the only form an EM field can take! As the frequency increases above that of conventional radio, we encounter *microwaves*. As the frequency keeps going up, we come to *infrared* (IR) radiation, also imprecisely called "heat rays." After that comes *visible light*, then *ultraviolet* (UV) radiation, then *X rays*, and then *gamma rays*.

In the opposite sense, EM fields can exist at frequencies far below those of conventional radio signals. In theory, an EM wave can go through one complete cycle every hour, day, year, thousand years, or million years. Some astronomers suspect that stars and galaxies generate EM fields with periods of years, centuries, or millennia.

THE EM WAVELENGTH SCALE

In order to illustrate the range of EM wavelengths, we can use a *logarithmic nomograph*. The logarithmic scale is needed because the range is so great that a linear scale is impractical. The left-hand portion of Fig. 11-6 is a logarithmic nomograph that shows wavelengths from 10^8 m all the way down to 10^{-12} m. Each division, in the direction of shorter wavelength, represents a tenfold decrease, known as an *order of magnitude*. Utility AC is near the top of this scale; the wavelength of 60-Hz AC in free space is

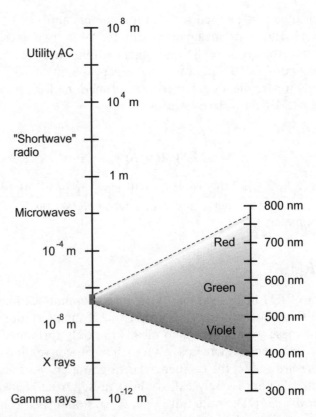

Fig. 11-6.   At left, a logarithmic nomograph of the most commonly observed part of the EM spectrum. At right, a magnified view of the visible-light range within this spectrum.

quite long. The gamma (γ) rays are denoted approximately at the bottom; their EM wavelengths are tiny.

Visible light, as can be seen from the left-hand nomograph, takes up only a tiny sliver of the whole range of EM wavelengths, known as the *EM spectrum*. In the right-hand scale, which is an expansion of the visible-light portion of the spectrum, the wavelengths are denoted in nanometers (nm).

⸛ **PROBLEM 11-3**

What is the wavelength of the 60 Hz EM field produced in free space by the AC in a common utility line?

✔ **SOLUTION 11-3**

Use the formula for wavelength in terms of frequency:

$$\lambda = 3.00 \times 10^8 / f$$
$$= 3.00 \times 10^8 / 60$$
$$= 5.0 \times 10^6 \text{m}$$

This is approximately 5,000,000 m or 5000 km, which is half the distance from the earth's equator to the north geographic pole as measured over the surface. That's approximately the distance between Maine and Ecuador!

⸛ **PROBLEM 11-4**

Suppose AC is passed through a wire conductor, and the frequency of the AC wave is variable. This gives rise to an EM field, the wavelength of which varies as the AC frequency varies. Illustrate, by means of a general graph, how the AC frequency and the free-space EM wavelength are related.

✔ **SOLUTION 11-4**

The frequency is inversely proportional to the wavelength. Figure 11-7 is an approximate graph of the free-space EM wavelength as a function of the AC frequency.

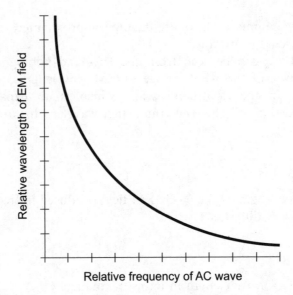

Fig. 11-7.   Illustration for Problem 11-4.

? **PROBLEM 11-5**

What is the frequency of the AC that produces an EM field with a wave-length of 75 m in free space? Express the frequency in hertz (Hz), kilohertz (kHz), and megahertz (MHz).

✔ **SOLUTION 11-5**

Use the formula given above for frequency in terms of wavelength, and plug in the numbers:

$$f = 3.00 \times 10^8 / \lambda$$
$$= (3.00 \times 10^8 / 75) \text{ Hz}$$
$$= (0.04 \times 10^8) \text{ Hz}$$
$$= (4 \times 10^6) \text{ Hz}$$
$$= 4,000,000 \text{ Hz}$$
$$= 4000 \text{ kHz}$$
$$= 4 \text{ MHz}$$

Electromagnetic Interference

Electromagnetic interference (EMI) is a phenomenon in which electronic devices upset each other's operation. In recent years, this problem has been getting worse because consumer electronic devices are proliferating, and they have become more susceptible to EMI. Wireless devices and systems are particularly vulnerable to this problem, both as "perpetrators" and as "victims."

COMPUTERS, HI-FI, AND TV

Much of the EMI that plagues consumer devices is the result of inferior equipment design. Faulty installation also contributes to the problem. A personal computer (PC) produces wideband radio-frequency (RF) energy in the form of an EM field. If the PC uses a CRT display, the trouble is compounded because these devices generate EM fields as a result of the activity of their deflection coils. The digital pulses in the central processing unit (CPU) and peripherals can also cause problems in some cases. The EM fields are radiated from unshielded interconnecting cables and power cords, because these cables and cords act as miniature transmitting antennas (Fig. 11-8A).

Computers, television (TV) receivers, and high-fidelity (hi-fi) sound equipment can malfunction because of strong RF fields, such as those from a nearby radio or a television transmitter. The problem is usually the result of an inadequate *RF shielding* in some part of the home-entertainment system or computer. In a computer system, cables and cords can act as receiving antennas (Fig. 11-8B), thereby letting RF energy into the CPU. This can cause the microprocessor to "go crazy." In high-fidelity (hi-fi) sound equipment, RF can get in by means of unshielded speaker wires, the power cord, the antenna and feed line, and unshielded cables between an amplifier and externals such as a compact disc (CD) player, as shown in Fig. 11-8C. In TV receiving installations, RF can enter through the power cord, the cable system, a satellite antenna system, and cords between the TV set and peripheral equipment such as a video disc recorder (Fig. 11-8D). Shielded cables, such as those shown in Fig. 11-8D between the TV set and the cable network, and between the TV set and the signal decoder, are far less susceptible to EMI.

In general, as the number of interconnecting cables in a home entertainment system increases, the system becomes more susceptible to

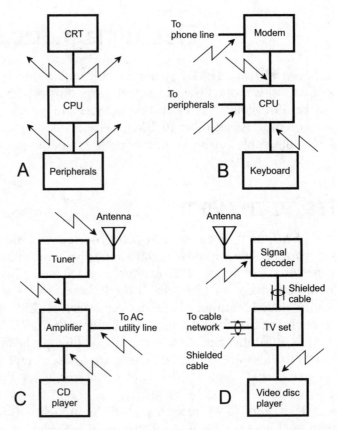

Fig. 11-8. At A, EMI radiated from a personal computer system. At B, a computer system can be susceptible to EMI. At C, a hi-fi system can pick up EM fields. At D, shielded cables offer partial protection against EMI in a home TV receiving system.

EMI. Also, as the interconnecting cables are made longer, the potential for trouble increases. It is wise to use as few connecting cords as possible, and to keep them as short as possible, in all home entertainment systems. If there is excess cord or cable and you don't want to cut it shorter, coil it up and tape it in place. A good electrical ground is important for hi-fi and TV systems. Besides offering some protection against possible electrocution, a proper grounding can reduce the susceptibility of the system to EMI.

IN HAM RADIO

If you're an amateur radio operator or a shortwave radio enthusiast, then you would have experienced, or will experience, EMI-related radio reception problems. If you are a radio ham with a sophisticated station, you might be blamed for interference to home entertainment equipment, whether it is technically your fault or not. Amateur radio operators have been blamed for things that would astound competent engineers. But there are certain circumstances in which the RF fields from amateur radio transmitters can be dangerous. Transmitting antennas should be located so humans cannot come into direct contact with them. At microwave frequencies, the RF fields emitted by high-power transmitters connected to directional antennas can cause deep burns at close range. The effects of long-term exposure to low-intensity RF fields is not well known, and is a subject of ongoing research.

When EMI originates from a broadcast station, a ham radio station, a CB radio station, a cell phone, or any other wireless transmitting device, it is called *radio-frequency interference* (RFI). This type of interference is rarely the fault of the radio transmitter, which is merely doing its job of generating RF energy. Nevertheless, RFI from ham radio equipment can often be eliminated by the amateur operator (even if it is not technically his or her fault) if the transmitter RF power output is reduced, the operating frequency is changed, or operating hours are chosen with consideration.

? **PROBLEM 11-6**

How can the arrangement shown in Fig. 11-8C be made less susceptible to RFI from nearby radio transmitters?

✔ **SOLUTION 11-6**

This illustration shows a hi-fi system in which none of the cords or cables are shielded. If all the interconnecting cables are shielded, the system will be less susceptible to RFI from nearby radio transmitters. The cable shields (usually a braid surrounding the internal conductor or conductors) should all be connected to the chassis of the amplifier. The chassis of the amplifier should be connected to a good earth ground.

EM versus ELF

Many electrical and electronic devices produce EM fields with wavelengths much longer than the wavelengths of radio and wireless signals. Such EM fields are caused by currents having *extremely low frequency* (ELF). That's how the term *ELF fields* originated.

WHAT ELF IS (AND ISN'T)

The *ELF spectrum* begins at frequencies of less than 1 Hz, and extends upward to approximately 3 kHz. This corresponds to wavelengths longer than 100 km. The most common ELF fields in today's environment have a frequency of 60 Hz, and are emitted by utility wires in the United States and many other countries. (In some countries it is 50 Hz.) The military has an ELF installation that is used to communicate with submarines. The ELF waves travel underground and underwater more efficiently than radio waves at higher frequencies.

The term *extremely-low-frequency radiation* is misleading. The words "extreme" and "radiation" suggest danger, and this has been exploited by fear-mongers. The media attention that has been given to ELF fields has led some people to purchase devices that supposedly offer personal protection from the "threat," but in fact do nothing at all.

An ELF field is not like a barrage of X rays or gamma rays, which can cause immediate sickness and rapid death, if received in large doses. Neither does ELF energy resemble UV, which has been linked to skin cancer, or IR, which can cause burns. An ELF field will not make anything radioactive. Some scientists suspect, nevertheless, that long-term exposure to high levels of ELF is correlated with an abnormally high incidence of certain health problems. This is a hotly debated topic and, like any such issue, has become politicized.

ELF AND COMPUTER DISPLAYS

An ELF source that has received much publicity is the cathode-ray tube (CRT) display, of the sort used in older desktop personal computers. (Actually, CRT displays produce EM energy at higher frequencies, not only at ELF.) In the CRT, images are created as electron beams strike a phosphor coating on the inside of the glass. The electrons change direction as they sweep from left to right, and from top to bottom, on the screen. The sweeping is caused by deflecting coils that steer the beam across the screen. The coils generate fluctuating magnetic fields that interact with the electrons, giving rise to EM fields at frequencies from a few hertz up to a few kilohertz.

Because of the positions of the coils, and the shapes of the fields surrounding them, more EM energy is emitted from the sides of a CRT display cabinet than from the front. If there's any health hazard, it is greatest for someone sitting off to the side of a CRT display, and least for someone looking at the screen from in front. The best "shielding" from EM energy is physical distance. This is especially true for people sitting next to (rather than in front of) a desktop computer display. The EM field dies off rapidly with distance from the display cabinet. Computer workstations in an office environment should be at least 1.5 m (about 5 ft) apart. You should keep at least 0.5 m (about 20 in.) away from the front of your own display. The display can be shut off when it's not in use.

Special CRT displays, designed to minimize the radiation of EM fields, are available. They are expensive, but they can offer peace of mind for people concerned about possible long-term health effects from exposure to EM fields. The newer liquid-crystal displays (LCDs) emit essentially no EM energy, and have replaced CRT displays in many computer systems. Most people who have seen a good LCD "fall in love" with it immediately because of its crisp images, appealing flat-panel design, reasonable weight, and small "desktop footprint." The largest LCDs are expensive, but they are coming down in price.

?   **PROBLEM 11-7**

What is the wavelength of an ELF field that results from AC at a frequency of 10 Hz? Express the answer in meters (m), and also in kilometers (km).

✔   **SOLUTION 11-7**

Use the formula stated earlier in this chapter for wavelength in terms of frequency:

$$\lambda = 3.00 \times 10^8 / f$$
$$= (3.00 \times 10^8 / 10) \text{ m}$$
$$= 3.00 \times 10^7 \text{ m}$$
$$= 30,000,000 \text{ m}$$
$$= 30,000 \text{ km}$$

That's three-quarters of the way around the world!

Quiz

This is an "open book" quiz. You may refer to the text in this chapter. A good score is 8 correct answers. Answers are in Appendix 1.

1. The deflecting coils in an electromagnetic CRT cause the electron beam to change direction because

    (a) the electrons are tiny magnets with only one pole, and they are either attracted to or repelled from the coils for that reason.
    (b) the flow of electrons in the beam constitutes current, which sets up a magnetic field that interacts with the magnetic field from the coils. The result is a magnetic force on the electrons.
    (c) the electrons in the coils produce an electrostatic repulsive force against the electrons in the moving beam.
    (d) the moving electrons cause the coils to become electrically charged, so one coil electrostatically attracts the electrons, while the other coil electrostatically repels them.

2. Which of the following statements about ELF fields is true?

    (a) They occur at frequencies lower than the frequencies of conventional radio waves.
    (b) They can make nearby objects radioactive.
    (c) They are not emitted by devices found in everyday life, but are only produced by the earth's magnetic field.
    (d) There is no way to effectively minimize their effects, if any, on people.

3. As the wavelength of an EM wave in free space increases, the frequency at which the electric field fluctuates

    (a) increases without limit.
    (b) increases toward a specific upper limit.
    (c) does not change.
    (d) decreases.

4. A galvanometer made with a magnetic compass and a coil of wire works because of the combined effects of

    (a)   the magnetic field around the coil and the magnetic field produced by the earth.
    (b)   the magnetic field around the compass needle and the magnetic field produced by the earth.
    (c)   the current through the coil and the current through the compass needle.
    (d)   the current through the compass needle and the current through the earth.

5. Which of the following phenomena does not produce an EM field?

    (a)   A fluctuating current in a wire.
    (b)   A variable charge gradient between two objects that are close to each other.
    (c)   A constant voltage on an electrode.
    (d)   High-frequency AC in a wire.

6. In free space, the product of the frequency and the wavelength of an EM field is equal to

    (a)   0.
    (b)   1.
    (c)   the speed of light.
    (d)   a quantity that varies depending on the frequency.

7. Fill in the blank to make the following sentence true: "The electric lines of flux in an EM field are oriented _____ the EM field travels through space."

    (a)   in the same direction as
    (b)   in the opposite direction from that which
    (c)   at right angles to the direction in which
    (d)   at random with respect to the direction in which

8. Fill in the blank to make the following sentence true: "The magnetic lines of flux in an EM field are oriented _____ the EM field travels through space."

    (a)   in the same direction as
    (b)   in the opposite direction from that which
    (c)   at right angles to the direction in which
    (d)   at random with respect to the direction in which

9. Suppose a magnetic compass is near a wire carrying 10 A DC, and the compass needle is affected by the current in the wire, the geomagnetic field, and no other factors. Suppose the needle is pointing exactly 15° to the east of the geomagnetic pole. Then the current in the wire drops to zero. The compass needle will move so it points

    (a) less than 15° to the east of the geomagnetic pole.
    (b) exactly 15° to the west of the geomagnetic pole.
    (c) less than 15° to the west of the geomagnetic pole.
    (d) directly at the geomagnetic pole.

10. Suppose a magnetic compass is near a wire carrying 10 A DC, and the compass needle is affected by the current in the wire, the geomagnetic field, and no other factors. Suppose the needle is pointing exactly 15° to the east of the geomagnetic pole. Then the DC in the wire changes to 5 A DC in the opposite direction. The compass needle will move so it points

    (a) less than 15° to the east of the geomagnetic pole.
    (b) exactly 15° to the west of the geomagnetic pole.
    (c) less than 15° to the west of the geomagnetic pole.
    (d) directly at the geomagnetic pole.

CHAPTER 12

Practical Magnetism

Magnetic substances and effects are involved in the workings of various devices and systems. In this chapter, we'll look at some common examples.

Electromagnets

When a rod made of ferromagnetic material, called a *magnetic core*, is placed inside a coil of wire (Fig. 12-1), the result is an electromagnet. The magnetic flux produced by the current temporarily magnetizes the core material. In addition, the magnetic lines of flux are concentrated in the core, so if the current is large, the field strength in and near the core can be considerable.

DC ELECTROMAGNETS

You can build a *DC electromagnet* by taking a large iron or a steel bolt and wrapping a couple of hundred turns of wire around it. These items are available in almost any hardware store. Be sure the bolt is made of ferromagnetic material, but don't use a permanent magnet. Ideally, the bolt should be at least 3/8 in. in diameter and several inches long. You must use insulated or

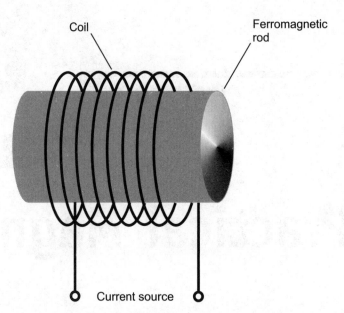

Fig. 12-1.   A simple electromagnet consists of a wire coil surrounding a rod made of ferromagnetic material.

enameled wire, preferably made of solid, soft-drawn copper. The so-called *bell wire* works well. Be sure all the wire turns go in the same direction. A 6 V lantern battery can provide plenty of current to operate the electromagnet. Even a small dry cell such as an AA alkaline cell can provide enough current to make a good demonstration.

When using a battery-powered electromagnet, don't leave the coil connected to the battery for more than a few seconds at a time. And beware: *Do not use an automotive battery for this experiment.* The near-short-circuit produced by an electromagnet can cause the acid from a lead-acid battery to boil out, and that acid can burn you. Clothes offer little, if any, protection (as I once learned from a painful and horrifying experience). If the acid gets into your eyes, it can blind you. Stick to the dime-store dry cells and batteries!

Direct-current electromagnets have defined north and south poles, just as do permanent magnets. The main difference is that an electromagnet can get much stronger than any permanent magnet. You should see evidence of this if you do the above experiment with a fresh lantern battery, a large bolt, and a coil with a lot of turns. Another difference between an electromagnet and a permanent magnet is the fact that, in an electromagnet, the magnetic field exists only as long as the coil carries current. When the power source is removed, the magnetic field collapses and disappears almost completely. A small amount of *residual magnetism* remains in the core, but this is much weaker than the magnetism generated when current flows in the coil.

AC ELECTROMAGNETS

Do you get the idea that the electromagnet can be made immensely powerful if, rather than using a lantern battery for the current source, you use utility AC instead? In theory, this is the case. But here is another warning: *Do not try to make an AC electromagnet by winding wire around a ferromagnetic core and then plugging the coil into a wall outlet!* This will overload common electrical circuits, can cause fires, and expose you to the risk of electrocution.

Some commercially manufactured AC electromagnets operate from household utility power, but they are designed with safety in mind. They are equipped with devices such as step-down transformers, current limiters, and circuit breakers. These magnets "really stick" to ferromagnetic objects. But they don't have constant north or south magnetic poles. The polarity of the magnetic field reverses every time the direction of the current reverses. There are 120 polarity reversals, or 60 complete north-south-north polarity cycles, every second (Fig. 12-2).

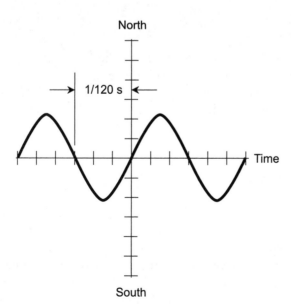

Fig. 12-2. The polarity of an AC electromagnet reverses every 1/120 of a second when the AC frequency is 60 Hz.

? **PROBLEM 12-1**

Suppose the frequency of the current applied to a well-designed commercial AC electromagnet is 50 Hz, as it is in some countries, instead of the nominal 60 Hz commonly used in the United States. What will happen to the interaction between the alternating magnetic field and a nearby ferromagnetic substance such as iron?

✔ **SOLUTION 12-1**

Assuming there is no change in the core material, the situation will be essentially the same as is the case at 60 Hz. An AC electromagnet will function just as well at 50 Hz as at 60 Hz.

? **PROBLEM 12-2**

Suppose the frequency of the current applied to an AC electromagnet is increased far above 60 Hz. Imagine that you have a variable-frequency AC electrical source of constant voltage, capable of delivering constant current. You increase the frequency to 600 Hz, then 6 kHz, then 60 kHz, 600 kHz, 6 MHz, 60 MHz, 600 MHz, and all the way up to 6 GHz! What will happen to the magnetic field as the frequency increases?

✔ **SOLUTION 12-2**

The electromagnet will work normally as the frequency increases, until the coil-and-core combination begins to impede the flow of AC. Then the magnetic field strength will get smaller and smaller, gradually tapering off to near zero as the frequency continues to rise.

Magnetic Materials

Some substances cause magnetic lines of flux to bunch closer together than is the case in *free space* (dry air or a vacuum). A few materials cause the lines of flux to spread farther apart than is the case in free space. The first kind of material is, as we have learned, called ferromagnetic. Substances of this type are "magnetizable." The other kind of material is called *diamagnetic*. Wax, dry wood, bismuth, and silver are examples of substances that decrease magnetic flux density as compared to the flux density in free space. The magnetic characteristics of a particular substance can be quantified in two important but independent ways: *permeability* and *retentivity*.

PERMEABILITY

Permeability, symbolized by the lowercase Greek mu (μ), is measured on a scale relative to free space. In theory, permeability values can range from 0 to "infinity" (arbitrarily large values). A perfect vacuum is assigned, by convention, a permeability of 1. If current is forced through a wire loop or coil in dry air, then the flux density in and around the coil is about the same as it would be in a vacuum. The permeability of dry air is approximately equal to 1.

If a core made of a ferromagnetic material such as iron, nickel, or steel is placed inside a coil, the flux density increases compared with the flux density in free space. Many common substances make the flux density thousands of times greater than it is in dry air or a vacuum. If certain special metallic alloys are used as the core materials in electromagnets, the flux density, and therefore the local strength of the field, can be increased by a factor of up to about a million. So, once in awhile you'll hear or read about a substance in which $\mu = 10^6$!

Diamagnetic materials, in contrast, dilate the magnetic lines of flux passing through them, and in that sense they are "antimagnetic." They have permeability values less than 1, but only a little less under ordinary circumstances. In most situations, diamagnetic substances are used to keep magnets or ferromagnetic objects physically separated while minimizing the interaction between them. Under extreme low-temperature conditions, some electrical conductors lose all their resistance. When this happens, they attain a theoretical permeability of 0. Such a material expels magnetic flux. That means a magnetic field can't exist inside it at all! These materials can be used to produce *magnetic levitation*, as we'll see later in this chapter.

Table 12-1 lists some common materials and their approximate permeability values or ranges.

RETENTIVITY AND CORE SATURATION

When a ferromagnetic substance such as iron or steel is subjected to a magnetic field by enclosing it in a wire coil carrying a high current, there is always some residual magnetism left when the current stops flowing in the coil. *Retentivity*, also called *remanence*, is a measure of how well a

Table 12-1   Approximate permeability values or ranges for some common substances.

Substance	Approximate permeability
Air, dry	1
Aluminum	Slightly more than 1
Bismuth	Slightly less than 1
Cobalt	60–70
Ferrite	100–3000
Iron	60–100
Nickel	50–60
Permalloy	3000–30,000
Silver	Slightly less than 1
Steel	300–600
Super permalloys	100,000–1,000,000
Vacuum	1
Wax	Slightly less than 1
Wood, dry	Slightly less than 1

substance can "memorize" a magnetic field imposed on it, and thereby become a permanent magnet.

In the real world, if you make an electromagnet with a core material, there is a limit to the flux density that can be generated in that core. As the current in the coil increases, the flux density inside the core goes up in proportion for a while. Beyond a certain point, the flux density levels off, and further increases in current do not produce any further increase in the flux density. This condition is called *core saturation*. When we determine retentivity for a material, we must compare the flux density when the material is saturated to the flux density when the magnetomotive force has been removed.

Retentivity can be expressed as a percentage. Suppose that for a particular substance, the maximum possible flux density, called the *core-saturation flux density*, is x tesla or gauss. Suppose the flux density in the core diminishes to y tesla or gauss when the current is removed from the coil. Then y is a fraction of x. The retentivity, $B_{r\%}$, of the material is given by the following formula:

$$B_{r\%} = (100y/x)\%$$

Imagine that a metal rod can be magnetized to a flux density of up to 1000 G when it is enclosed by a coil carrying an electric current. Imagine that this is the maximum possible flux density that the rod can be forced to have. For any substance, there is always such a maximum; further increasing the current in the wire does not make the flux density within the core any greater. Now suppose the current is shut off, and the flux density in the rod drops to 20 G. Then the retentivity, $B_{r\%}$, is:

$$B_{r\%} = (100 \times 20/1000)\%$$
$$= (100 \times 0.02)\%$$
$$= 2\%$$

Certain ferromagnetic substances have high retentivity, and are excellent for making permanent magnets. Other ferromagnetic materials have low retentivity. They can work well as the cores of electromagnets, but they do not make good permanent magnets.

Sometimes it is desirable to have a substance with good ferromagnetic properties, but low retentivity. This is the case when you want to have an electromagnet that will operate from DC, so that it maintains a constant polarity, but that loses its magnetism when the current is shut off. If a ferromagnetic substance has low retentivity, it works well as the core for an AC electromagnet, because the polarity is easy to switch. If the retentivity is high, the material is "magnetically sluggish" and cannot follow rapid current reversals in the coil. Thus, substances with high retentivity do not work well in AC electromagnets.

? **PROBLEM 12-3**

Suppose a metal rod is surrounded by a coil, and the magnetic flux density can be made as great as 0.5 T; further increases in current cause no further increase in the flux density inside the core. Then the current is removed; the flux density drops to 500 G. What is the retentivity of this core material?

✔ **SOLUTION 12-3**

First, convert both the flux density figures to the same units. Remember that $1 \text{ T} = 10^4 \text{ G}$. Therefore, the flux density is $0.5 \times 10^4 = 5000 \text{ G}$ with the current. We are told it is 500 G without the current. "Plugging in" these numbers gives us this:

$$
\begin{aligned}
B_{r\%} &= (100 \times 500/5000)\% \\
&= (100 \times 0.1)\% \\
&= 10\%
\end{aligned}
$$

? **PROBLEM 12-4**

What is the retentivity, in general, of a diamagnetic material such as wax or dry wood? How about pure, dry air?

✔ **SOLUTION 12-4**

Any material that has permeability of 1 or less does not concentrate magnetic flux at all. If the core inside a coil is made of a non-ferromagnetic material, it doesn't retain any magnetic flux after the current is removed from the coil. Therefore, the retentivity is zero or 0%.

PERMANENT MAGNETS

Any ferromagnetic material, or substance whose atoms can be permanently aligned, can be made into a permanent magnet. These are the magnets you played with as a child (and maybe still play with when you use them to stick notes to your refrigerator door). Some alloys can be made into stronger permanent magnets than others.

Permanent magnets are best made from materials with high retentivity. They are made by using the material as the core of an electromagnet for an extended period of time. Weak permanent magnets can be made using other permanent magnets. If you want to magnetize a screwdriver so it holds onto screws, stroke the shaft of the screwdriver with the end of a bar magnet several dozen times. But take note: Once you have magnetized a tool, it is just about impossible to completely demagnetize it.

FLUX DENSITY INSIDE A LONG COIL

Suppose you have a long coil of wire, commonly known as a *solenoid*, with n turns, and whose length in meters is s. Suppose this coil carries a direct current of I amperes, and has a core whose permeability is μ. The flux density in teslas, B, inside the core, assuming that it is not in a state of saturation, can be found using this formula:

$$B = 4\pi \times 10^{-7}(\mu nI/s)$$

A good approximation is:

$$B = 1.2566 \times 10^{-6}(\mu nI/s)$$

? **PROBLEM 12-5**

Consider a DC electromagnet that carries a certain current. It measures 20 cm long, and has 100 turns of wire. The flux density in the core, which is not in a state of saturation, is 20 G. The permeability of the core material is 100. What is the current in the wire?

✔ **SOLUTION 12-5**

First, convert units to the correct form, as necessary. The length, s, is 20 cm, equivalent to 0.2 m. The flux density, B, is 20 G, equivalent to 0.002 T. Next, rearrange the above formula to solve for I:

$$B = 1.2566 \times 10^{-6}(\mu nI/s)$$
$$I = 7.9580 \times 10^{5}(sB/\mu n)$$

Finally, plug in the numbers:

$$I = 7.9580 \times 10^{5}(0.2 \times 0.002)/(100 \times 100)$$
$$= 7.9580 \times 10^{5} \times 4.0 \times 10^{-8}$$
$$= 0.031832\,\text{A} = 31.832\,\text{mA}$$

This can be rounded off to 32 mA.

Magnetic Devices

Here are some examples of devices that make use of permanent magnets or electromagnets. They are all *electromechanical transducers* because they convert electrical energy to mechanical motion, or vice-versa.

A BELL RINGER

Figure 12-3 is a simplified diagram of a bell ringer. Its solenoid is an electromagnet. The core has a hollow region in the center, along its axis, through which a steel rod passes. The coil has many turns of wire, so the electromagnet is powerful if a substantial current passes through the coil.

When there is no current flowing in the coil, the rod is held down by gravity. When a pulse of current is first applied to the coil, the rod is pulled

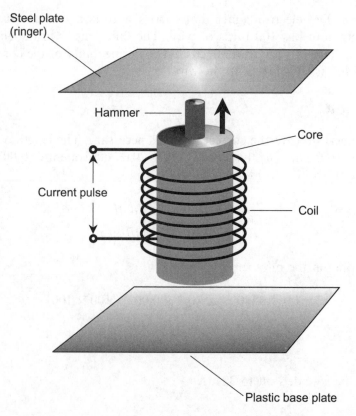

Fig. 12-3.   Functional diagram of a solenoid that works as a bell ringer.

forcibly upward. The magnetic field "tries" to align the ends of the rod with the ends of the core (and if the current were left on indefinitely, this would happen). But the current pulse is brief, and by the time the rod is lined up with the core, the current pulse has ended and the magnetic field has disappeared. The upward momentum of the rod causes it to continue upward, strike the ringer plate, and bounce down again. The ringer plate reverberates: "Dingggg!" The rod returns to its resting position and awaits the next current pulse.

Some office telephones are equipped with ringers that produce intermittent chimes at intervals of 2 or 3 seconds, rather than the conventional jangling, tweeting, or buzzing noises emitted by other types of phone sets. The intermittent chimes are less irritating to some people than other attention-demanding signals.

THE RELAY

In some electronic devices, it is inconvenient to place a switch at the optimum location. For example, you might want to switch a communications line from one branch to another from a long distance away. In wireless transmitters, some of the wiring carries high-frequency AC that must be kept within certain parts of the circuit, and not routed out to the front panel for switching. A *relay* makes use of a solenoid to allow remote-control switching.

Figure 12-4A is a functional drawing of a relay, and Fig. 12-4B is the schematic symbol for the same relay. This particular unit is called a *single-pole double-throw (SPDT) relay* because it can switch a single terminal between two different circuits. The movable lever, called the *armature*, is held to one side by a spring or by its own tension when there is no current flowing through the electromagnet. Under these conditions, terminal X is connected to Y, but not to Z. When sufficient current is applied to the coil, the armature is pulled down by the magnetic force exerted on a ferromagnetic disk by the solenoid. This disconnects the terminal X from Y, and connects X to Z.

There are numerous types of relays, used for different purposes. Some are meant for use with DC, and others are intended for switching AC. Some can work with either AC or DC. A *normally closed relay* completes the circuit when there is no current flowing in its solenoid, and breaks the circuit when current flows. A *normally open relay* is just the opposite. ("Normal" in this sense means no current in the solenoid.) The relay shown in Fig. 12-4 can be used either as a normally open or closed relay, depending on which contacts are selected. It can also be used to switch a line between two different circuits.

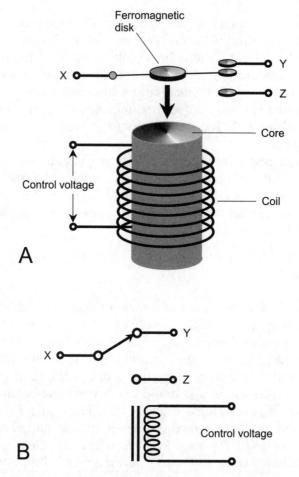

Fig. 12-4. At A, a functional diagram of a relay. At B, the schematic symbol.

Nowadays, relays are used only in circuits and systems carrying high currents or voltages, or where exceptional physical ruggedness is required. In most applications, electronic semiconductor switches are preferred. Semiconductor devices such as diodes and transistors have no moving parts, so they can perform many more switching operations per second than relays do.

THE DC MOTOR

Magnetic fields can be harnessed to do mechanical work. The device that converts DC energy into rotating mechanical energy is a *DC motor*.

Some motors are microscopic in size, and others are as big as a house; there are motors of all sizes in between. Some tiny motors are being considered for use in medical devices that circulate in the bloodstream or are installed in body organs. Others can pull a train at hundreds of kilometers per hour.

In a DC motor, the source of electricity is connected to a set of coils, producing magnetic fields. The attraction between the opposite poles, and the repulsion between like poles, is switched in such a way that a constant torque, or rotational force, results. The greater the current that flows in the coils, the stronger the torque, and the more electrical energy it needs to turn the motor shaft.

Figure 12-5 is a simplified functional diagram of a DC motor. The *armature coil* rotates with the motor shaft. A pair of coils called the *field coil* is stationary. In some motors, the field coil is replaced by a pair of permanent magnets. The current direction in the armature coil is reversed every half-rotation by the *commutator*. This keeps the force going in the same sense, either clockwise or counterclockwise. The shaft is carried along by its own rotational momentum, so it doesn't stop at the instants in time when the current direction is switching.

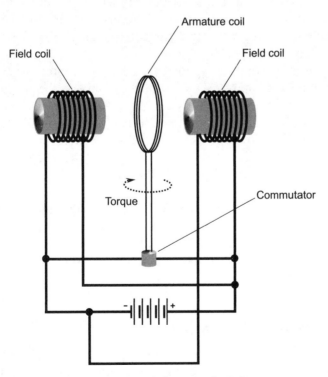

Fig. 12-5.   Functional diagram of a DC motor.

THE STEPPER MOTOR

A *stepper motor* turns in small increments, rather than continuously. The *step angle*, or extent of each turn, varies depending on the particular motor. It can range from less than 1° of arc to a quarter of a circle (90°). A stepper motor turns through its prescribed step angle and then stops, even if the current is maintained. In fact, when a stepper motor is stopped with a current going through its coils, the shaft resists applied torque.

Conventional motors run at hundreds, or even thousands, of revolutions per minute (rpm). A stepper motor usually runs at less than 180 rpm, and often much less. In a conventional motor, the torque increases as the motor runs faster. But in a stepper motor, the torque decreases as the motor speed increases. A stepper motor has the most turning power when it is running at slow speed. When it is stopped, it tends to stay in place and resist applied rotational forces. It has, in effect, electromagnetic brakes!

The most common stepper motors are of two types. A *2-phase stepper motor* has 2 coils, called *phases*, controlled by 4 wires. A *4-phase stepper motor* has 4 phases and 8 wires. The motors are advanced incrementally (stepped) by applying currents to the phases in a controlled sequence. Figure 12-6 shows schematic diagrams of 2-phase (A) and 4-phase (B) stepper motors. Table 12-2 shows control-current sequences for 2-phase (A) and 4-phase (B) stepper motors.

When a pulsed current is supplied to a stepper motor, with the current rotating through the phases, as shown in the tables, the motor rotates in increments, one step for each pulse. In this way, a precise speed is maintained. Because of the braking effect, this speed is constant for a wide range of mechanical loads (turning resistances). This is in contrast to a conventional DC motor, which tends to slow down when the mechanical load increases, or to speed up when the mechanical load decreases or is removed.

Stepper motors can be controlled using microcomputers. Several stepper motors, all under the control of a single microcomputer, are used in robotic devices. Stepper motors are especially well-suited for *point-to-point motion*, in which the position of a robotic device can be precisely changed at a programmable speed.

THE SELSYN AND SYNCHRO

A *selsyn* is an indicating device that uses stepper motors to show the direction in which an object is pointing. There is a transmitting motor connected to the

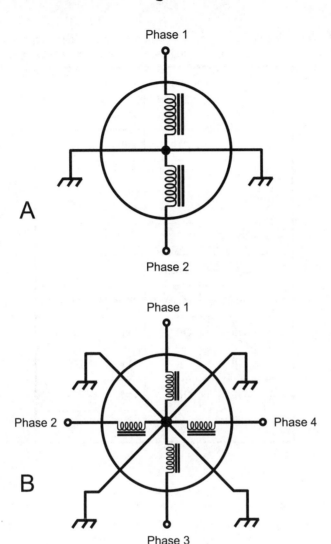

Fig. 12-6. At A, schematic symbol for a 2-phase stepper motor. At B, schematic symbol for a 4-phase stepper motor.

movable object, and a receiving motor at a convenient location. The transmitting motor senses the orientation of the object, and the receiving motor, equipped with a compass-like indicating needle, displays this orientation at a convenient location. A common application of the selsyn is as a direction indicator for a rotatable communications antenna.

In a selsyn, the indicator rotates the same number of degrees as the moving device. A selsyn designed to indicate the azimuth, or a horizontal compass

Table 12-2 Operation of a 2-phase stepper motor and a 4-phase stepper motor. Read down for clockwise rotation. Read up for counterclockwise rotation.

A. 2-phase stepper motor

Step	Phase 1	Phase 2
1	Off	Off
2	On	Off
3	On	On
4	Off	On

B. 4-phase stepper motor

Step	Phase 1	Phase 2	Phase 3	Phase 4
1	On	Off	On	Off
2	Off	On	On	Off
3	Off	On	Off	On
4	On	Off	Off	On

bearing, can rotate through a range of 360°. A selsyn designed to indicate elevation, or the angle with respect to the horizon, can rotate through a range of 90° for use on the earth's surface, or 180° for use in space.

A *synchro* is a special type of stepper motor, used for remote control of mechanical devices. A synchro consists of a generator and a receiver motor. As the shaft of the generator is turned by a control operator, the shaft of the receiver motor follows along exactly. A synchro is like a powerful selsyn in reverse. Synchros have many uses, especially in robotics. They are well-suited to fine motion, and also to robotic *teleoperation* (remote control). A selsyn can be used to remotely indicate the condition of a synchro.

Some synchro devices are programmable. The operator inputs a set of numbers into the synchro generator, and the receiver changes position accordingly. Computers allow sequences of movements to be programmed.

THE ELECTRIC GENERATOR

An *electric generator* is constructed somewhat like a conventional motor, although it functions in the opposite sense. Some generators can also operate as motors; such devices are called *motor/generators*. The operating principle of the generator is outlined in Chapter 6.

THE FLUXGATE MAGNETOMETER

A *fluxgate magnetometer* is a computerized mobile-robot guidance system that uses magnetic fields to tell a robot where it is located and how it is oriented. Navigation within a defined area can be carried out by having the *robot controller*, or microcomputer on board the robot, constantly analyze the orientation and intensity of the lines of flux generated by fixed electromagnets. Figure 12-7 shows a simplified hypothetical scenario, with two fixed electromagnets and a robot in the magnetic field produced by them. In this case, opposite magnetic poles (north and south) face each other, giving the flux field the characteristic pole-to-pole bar-magnet shape.

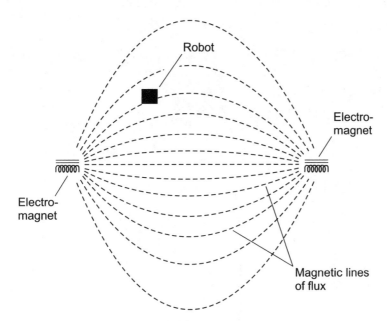

Fig. 12-7.   A fluxgate magnetometer makes it possible for a mobile robot to determine its location, based on the strength and orientation of the flux lines produced by fixed electromagnets.

For each point in the robot work environment, the magnetic flux has a unique orientation and intensity. There is a one-to-one correspondence between magnetic flux conditions and each point within the work environment. The robot controller is programmed to "know" this relation precisely for all points. This allows the robot controller to determine its location and orientation at all times. If the data is updated at frequent intervals, the velocity (speed and direction) at which the robot moves can also be determined.

THE DYNAMIC TRANSDUCER

A *dynamic transducer* is a coil-and-magnet device that converts mechanical vibration into electrical currents, or vice-versa. The most common examples are the *dynamic microphone* and the *dynamic speaker*.

Figure 12-8 is a functional diagram of a dynamic transducer. A thin, flexible, ferromagnetic *diaphragm* is attached to a permanent magnet that is mounted so it can move back and forth rapidly along its axis. The magnet is surrounded by a coil of wire.

When the transducer functions as a microphone, acoustic compression waves in the air cause the diaphragm to move. This moves the magnet along

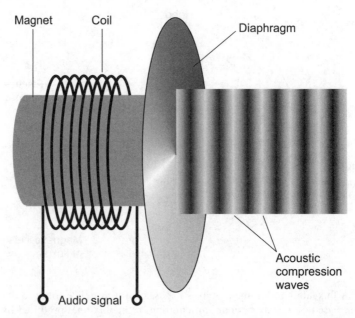

Fig. 12-8. Functional diagram of a dynamic transducer such as a microphone or speaker.

its axis, over a tiny distance, in a complex pattern that follows the acoustic waves. The back-and-forth motion of the magnet causes fluctuations in the magnetic field within the coil. The result is AC output from the coil, having the same waveform as the sound waves that strike the diaphragm.

When the transducer functions as a speaker, an audio-frequency AC signal is applied to the coil of wire. This creates a magnetic field that produces a back-and-forth force on the permanent magnet. This causes the magnet to move along its axis, over a tiny distance, pushing the diaphragm in and out. The motion of the diaphragm produces acoustic compression waves in the surrounding air.

[?] **PROBLEM 12-6**

What would happen if DC, rather than audio-frequency AC, were applied to the coil of a dynamic speaker?

[✔] **SOLUTION 12-6**

A magnetic field would be produced within the coil of wire. This would pull the permanent magnet either in or out, depending on the direction of the current. The diaphragm would therefore be "sucked in" or "pushed out," but it would not vibrate. No sound would be heard from the speaker, except for a click when the DC was first applied, and another click when the DC was removed. It is a bad idea to apply DC to a dynamic transducer, even in combination with an AC signal, because this can interfere with normal operation.

[?] **PROBLEM 12-7**

Can the earth's magnetic field be used, rather than an artificially generated magnetic field, in a fluxgate magnetometer system for a mobile robot?

[✔] **SOLUTION 12-7**

Yes. When this is done, the robot's work environment becomes, in effect, the entire surface of the earth, except for locations near the geomagnetic poles, or places where the earth's magnetic field cannot be sensed. But the robot will be misguided, if an artificially generated magnetic field is present in addition to the earth's magnetic field, unless the robot is programmed to correct for interaction between the two magnetic fields. Similar "confusion" can occur

for robots operating in concrete-and-steel buildings or in magnetically shielded enclosures.

Magnetic Media

The term *magnetic media* refers to hardware devices and systems that allow the storage and recovery of data in the form of magnetic fields. In computers, the most common magnetic medium is the *hard disk*, also called the *hard drive*. The *magnetic diskette*, once popular for external data storage, is on the way to obsolescence, having being largely replaced by optical media such as *compact disc* (*CD*) and electronic memory modules. *Magnetic tape* is sometimes used to store large amounts of data for long periods of time. In audio and video recording, magnetic tape is also used sometimes, although this medium being replaced by the *digital video disc* (*DVD*), which is similar to the CD but has more storage capacity.

HOW IT WORKS

In magnetic media, there are millions of tiny ferromagnetic particles attached to the surface of a tape or a disk. Each particle can be magnetized and demagnetized repeatedly. These particles can maintain a given state of magnetization for a period of years. The magnetic polarity can be in either direction (north pole facing up, or south pole facing up). Early in the 20th century, audio engineers noticed this property and saw its potential for recording sound. The result was the *audio tape recorder*. Around the middle of the century, engineers adapted audio tapes for use in the first computers. Later, flat disks were used instead, because they allow faster data access.

ADVANTAGES

Magnetic data can be erased and overwritten thousands of times. Magnetic media are a form of *non-volatile data storage*. That means they don't require a constant source of power to maintain the information content. You can switch a computer or tape recorder off, and the disk or tape will keep the information you've stored on it. You can put away a computer diskette, and then bring it out months or years later and find all the data unchanged.

Magnetic media have fairly high density. A typical computer diskette can hold 1.44 megabytes (MB), the equivalent of a long novel. A computer hard drive can hold many thousands of megabytes. Magnetic audio tapes, used in analog systems, typically play for 15 to 60 minutes per side, depending on the tape length and thickness.

LIMITATIONS

Magnetic media are heat-sensitive. If the temperature is too high, the ferromagnetic atoms move around so fast that their electron orbits rapidly lose their alignment. This effect occurs no matter what the temperature, given enough time, but extreme heat makes it happen fast enough that it can cause problems. Magnetic disks and tapes should be stored in a cool place. They should never be left in direct sunlight, or inside an enclosed car or truck during warm, sunny weather.

Magnetic media don't last forever, even if you provide ideal storage conditions. It's a good idea to renew everything (copy all your archives to new media) annually for computer disks, and every three to five years for tapes.

Magnetic media are sensitive to strong magnetic fields. Hence, keep disks and tapes away from permanent magnets, electromagnets, and from anything that generates magnetic fields. Loudspeakers, headphones, microphones, and the back ends of cathode-ray-tube (CRT) monitors are surrounded by magnetic fields. Common sense can go a long way here. Don't carry your computer diskettes in a handbag or briefcase along with magnets!

MAGNETIC DISKS AND HARD DRIVES

Magnetic disks have all the advantages and limitations of magnetic media in general. Disks enjoy one special advantage over their magnetic tape counterparts: data can be written onto, and read from, a disk much faster than to or from a tape. No two data elements are ever separated by more than the diameter of a disk (a few centimeters), while two different elements can be separated by the entire length of a tape (up to hundreds of meters).

You'll hear several terms when people talk about computer hard drives. A *platter* is one of the individual rigid disks within the system. There are several platters in the assembly. A *track* is one of the many concentric, circular paths on the disk surface, along which data is written. A *sector* is an arc-shaped portion of a track. A *cylinder* is the set of equal-diameter tracks on all the platters. The *access time*, also called the *read time*, is the length of time it takes to bring up files from the drive. The *storage time*, also called the *write time*, is the length of time it takes to save a file or a set of files to the drive. The *storage capacity* is the maximum amount of data that the drive can hold.

MAGNETIC TAPE

Magnetic tape is commonly used for storing sound, video, and digital information. It is available in various thicknesses and widths for different applications. The ferromagnetic particles are attached to a long, thin, flexible, stretch-resistant strip or *substrate* made of strong plastic. In some cases, non-ferromagnetic metal is used instead of plastic. Tapes from the "really old days" (circa 1940s) had substrates made of paper! A magnetic field, produced by a *recording head*, causes polarization of the particles. As the field fluctuates in intensity, the polarization of the particles varies in strength and alternates back and forth. When the tape is played back, the magnetic fields surrounding the individual particles produce current changes in the *playback head*, also called the *pickup head*.

For sound and computer-data recording, magnetic tape is available in cassette form or reel-to-reel form. The tape thickness can vary. Thicker tapes have better resistance to stretching, although the recording time, for a given length of tape, is proportionately shorter than with thin tape.

Magnetic tape provides a convenient and compact medium for long-term storage of information. But certain precautions must be observed. The tape must be kept clean and free from grease. Magnetic tapes should be kept at a reasonable temperature and humidity, and should not be subjected to magnetic fields.

You can't open files from a magnetic computer tape as fast as you can from a disk. Nevertheless, tapes can be useful for archiving computer data. The contents of a hard drive can be backed up on a cassette with a device called a *tape drive*.

A MAGNETIC TAPE RECORDER

For illustrative purposes, let's look at the operation of an old-fashioned analog tape recorder. All such devices act as transducers between acoustic, digital, or video signals and variable magnetic fields. A simplified rendition of the recording/playback apparatus in a typical audio tape recorder, of the sort you can still buy today at an electronics store or department store, is shown in Fig. 12-9.

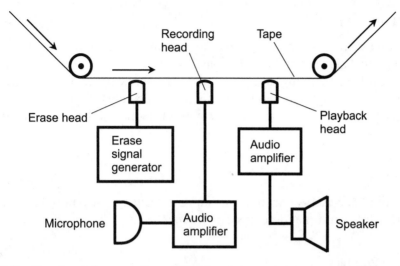

Fig. 12-9.   Functional diagram of an old-fashioned magnetic tape recorder. Illustration for Quiz Question 6.

In the *record mode*, the tape moves past the *erase head* before anything is recorded. If the tape is not blank (that is, if magnetic impulses already exist on it), the erase head removes these before anything else is recorded. This prevents *doubling*, which is the simultaneous presence of two programs on the tape. The *recording head* is an electromagnet that generates a fluctuating magnetic field, whose instantaneous flux density is proportional to the instantaneous level of the audio input signal. This magnetizes the tape in a pattern that duplicates the waveform of the signal.

In the *playback mode*, the erase head and the recording head are not activated, so they and their associated circuits become irrelevant. The *playback head* acts as a magnetic-field detector. As the tape moves past, the playback head is exposed to a fluctuating magnetic field whose waveform is identical to the waveform that was produced by the recording head when the recording was made. This magnetic field induces AC in the playback head. The AC is amplified and delivered to a speaker, headset, or any other output device.

? **PROBLEM 12-8**

Why is it recommended that magnetic tapes and disks be stored in places where the temperature doesn't rise too high?

✔ **SOLUTION 12-8**

There are two reasons for this. First, high temperatures reduce the lifetime of the magnetic fields around the tiny particles on a disk or tape. A well-known method of demagnetizing an object (such as a screwdriver or a wrench) is to heat it. Second, high temperatures can cause the substrate of a disk, and in particular the substrate of a tape, to soften. This makes the substrate stretch, warp, or shrivel.

? **PROBLEM 12-9**

What about the X-ray machines at airport security checkpoints? Can they erase the data on a magnetic disk or a tape?

✔ **SOLUTION 12-9**

X rays, by themselves, do not affect any magnetic disks or tapes, or the data on them. (However, X rays may erase or damage data on certain types of computer memory chips.) Walk-through and portable metal detectors generate magnetic fields, but there is a debate as to whether or not these fields are strong enough to affect the data on a magnetic disk or a tape. If you are in doubt, you can hand your tapes or diskettes to the attendants and let them pass them to you after you have walked through the metal detector. You can also send them by the old-fashioned post or by courier to your destination. Another solution is to store data on optical media such as CDs, which are immune to both the X rays and the magnetic fields. In any case, keep redundant backups of important data at multiple locations in case of loss, fire, theft, or other unfortunate events!

Magnetic Levitation

The next time you have access to a couple of powerful permanent magnets, hold them so their like poles are close together (north-to-north or south-to-south). As you push them closer to each other, feel the way they push back against you. Do you remember the first time you observed this? It's no wonder that magnets gave rise to visions of anti-gravity devices in the minds of early scientists!

AN EXPERIMENT

If you try to build a levitation device using plain magnets, you will soon become frustrated. Imagine that you have an access to a large number of small, pellet-shaped permanent magnets. Suppose you glue a couple of dozen of these magnets in a matrix all over the inside surface of a plastic mixing bowl, with the north poles facing upward. This produces a large magnet with a concave, north-pole surface. Suppose you anchor this bowl to a tabletop, and then take a single pellet-shaped magnet and hold it with its north pole facing downward, right over the center of the mixing bowl, as shown in Fig. 12-10A. As soon as you let go of the single magnet, it will flip over and—*Whack!*—it will stick to one of the magnets that is glued inside the bowl.

Now suppose that you take another couple of dozen magnets and glue them to another mixing bowl of the same shape, but only 2/3 or 3/4 as big as the first bowl. You glue the magnets all over the outside of the smaller bowl, with the north poles facing away from the bowl. This produces a large magnet with a convex north-pole surface. Suppose you try to set this bowl down inside the first one, as shown in Fig. 12-10B. Now, it is tempting to believe that the top "bowl magnet" will hover above the bottom one. But it won't.

EARNSHAW'S THEOREM

Magnetic levitation cannot be achieved with a set of static (non-moving and non-rotating) permanent magnets. There is always instability in such a system, and this instability is magnified by the slightest disturbance, ultimately causing the ensemble to crash. This fact was mathematically proven by a scientist named *Samuel Earnshaw* in the 1800s, and it has come to be known as *Earnshaw's Theorem*.

Despite the pessimistic conclusion of this theorem, it is possible to obtain magnetic levitation. This is because Earnshaw's Theorem is based on a narrow set of assumptions, and systems can be built that get around these constraints. Earnshaw's Theorem applies only to sets of *permanent* magnets in arrangements where there is *no relative motion*. In recent years, scientists have come up with dynamic (moving) systems of magnets that can produce levitation. Some such systems have been put to practical use.

FEEDBACK SYSTEMS

Consider the two-bowl scenario of Fig. 12-10B. If you had ever tried this experiment, you must have become frustrated. But suppose you build a

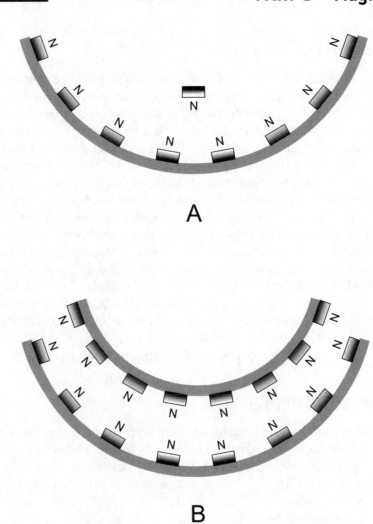

Fig. 12-10.   When you try to levitate a magnet above a set of other magnets as shown at A, the top magnet flips over and sticks to one of the others. Instability also occurs with two bowl-shaped magnetic structures, one above the other, as shown at B.

feedback system that keeps the upper bowl in alignment with the lower one? Feedback systems are common in all sorts of regulated devices, from the governors on motors to the oscillators in radio transmitters and receivers.

Here's a crude example of how an electromechanical feedback system can produce magnetic levitation with the two-bowl scheme. It has an electronic *position sensor* that produces an *error signal*, and a mechanical *position corrector* that operates, based on the error signals from the position sensor, to keep the upper bowl from drifting off center. As long as the upper bowl is exactly centered over the lower one (Fig. 12-11A), the position sensor produces no output signal. The upper bowl has a tendency to move sideways because of instability in the system (Fig. 12-11B). As soon as the bowl gets a

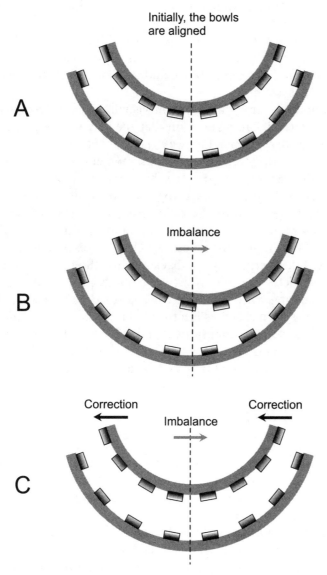

Fig. 12-11. A feedback system can keep two magnet-arrayed bowls centered, producing levitation. At A, the initial aligned situation. At B, an imbalance throws the upper bowl off center. At C, a correction force re-centers the upper bowl.

little off center, the position sensor produces data that describes the extent and direction of the displacement. The error data has two components: a *distance error signal* that gets stronger as the off-center displacement of the upper bowl increases, and a *direction error signal* that indicates the compass bearing (or azimuth) in which the upper bowl has drifted. These signals are sent to a microcomputer, which operates a mechanical device that produces the necessary amount of force, in exactly the right direction, to get the upper bowl back into alignment with the lower one (Fig. 12-11C). The mechanical device can be as crude as a set of fans, or as sophisticated as a second set of magnets mounted on the upper bowl, and a set of electromagnets around the periphery of the system.

Here's an analogy that will help you visualize how a feedback system of the sort just described can work. Consider the way you steer a car down the proper lane while driving on a highway. The car is inherently unstable. If you take your hands off the steering wheel for a moment, your car will veer out of its lane. But the car stays in its lane, because you constantly make minor course corrections. You and the car together constitute a feedback system. If the car starts to drift toward the right, your eye and brain sense this, and error data is sent to your hands. Your hands produce a slight counter-clockwise force on the steering wheel, which operates a system that turns the wheels a little bit, steering the car back toward the left. If the car starts to drift toward the left, you sense this, your hands produce a slight clockwise force on the steering wheel, and the car is steered back toward the right. A machine-vision system, some electronic circuits, and a microcomputer connected to the steering apparatus could substitute for you, and keep the car centered in its lane. Robot-driven cars have been designed, built, and tested. But don't expect to be tailgated by a robot vehicle any time soon.

DIAMAGNETIC MATERIALS

As we have seen, diamagnetic materials cause dilation of magnetic flux compared with the flux density in free space. These substances, under normal conditions, do not spread the lines of flux out nearly as much as ferro-magnetic materials concentrate them, but diamagnetic materials behave like "antimagnets."

A diamagnetic substance, such as dry wood or distilled water, repels either pole of a permanent magnet, just as a ferromagnetic substance such as iron attracts either pole. This is not noticed with ordinary magnets because they're not strong enough. When you, as a child, played with magnets, you didn't observe a force of repulsion against a hardwood floor, or against a piece of

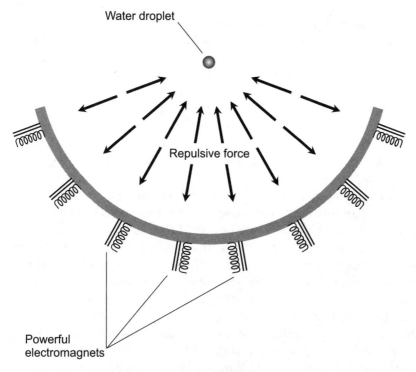

Fig. 12-12.   Levitation can be produced by the action of powerful electromagnets against a diamagnetic substance such as a droplet of water.

paper, or against a window pane. You didn't toss a permanent magnet at a pond and then watch it hover above the surface. It sank, of course, and water wasn't repelled from it, either. You noticed that some things attract magnets and other things don't, and you thought that was the end of the matter. But there is more to the story!

Repulsion between diamagnetic materials and magnets becomes significant if the magnets are strong enough. Levitation can occur if a lightweight diamagnetic object is placed inside a bowl-shaped container arrayed with specially designed, high-current electromagnets (Fig. 12-12). A small drop of distilled water, for example, can be suspended in midair. This has been done in lab experiments.

SUPERCONDUCTIVITY

Another phenomenon can be exploited to obtain magnetic levitation. At extremely low temperatures, approaching *absolute zero* (about −273° Celsius or −459° Fahrenheit), certain substances lose their electrical resistance. They become perfect conductors—or so nearly perfect that the practical difference between them and the ideal case is negligible. Such an electrical medium is called a *superconductor*, and the phenomenon of zero resistivity is called *superconductivity*. A small current in a superconducting loop of wire can circulate around and around, without growing noticeably weaker, for a long time. This is not a mere artifact of mathematical formulas; it has been done in laboratories. In theory, the current continues forever. In the real world, the current gradually dies down, but it remains significant for years!

Certain metals such as aluminum, and various alloys and compounds, exhibit superconductivity when they get cold enough. Other metals never behave in this manner, no matter how cold they get. In a substance that can function as a superconductor, there is a *critical temperature*, also called a *threshold temperature*, below which the effect takes place. Above that temperature, the substance behaves like a conventional electrical conductor, with small but measurable resistivity. If a superconducting substance is warmed to the critical temperature, its resistivity increases abruptly. Specially manufactured alloys called *cuprates* can superconduct at temperatures of more than 100° Celsius above absolute zero. However, as of this writing, no substance has yet been made to superconduct at room temperature.

Superconductors make magnetic levitation possible because magnetic flux is completely expelled. This phenomenon is called *perfect diamagnetism*. A more technical name for it is *Meissner Effect*, named after one of its discoverers, *Walter Meissner*, who first noticed it in the 1930s. A superconductor has a permeability that is equal to 0 for all practical purposes. Superconductors repel strong magnetic fields with considerable force. This is why superconductors have generated interest among engineers seeking to build machines, such as the so-called *maglev train*, that take advantage of magnetic levitation.

ROTATION

The arrangement shown in Fig. 12-10A doesn't work because the upper magnet flips over with the tiniest instability. In practice, it can't stay centered for more than a few milliseconds before a puff of air, a slight vibration in the table supporting the lower magnet array, or some other disturbance jostles it and—*Whack!*—its south pole is stuck to the north pole of one of the lower magnets.

There's a way to get an arrangement similar to that of Fig. 12-10A to work, and it is amazingly simple: set the upper magnet spinning. This makes it act like a stabilizing gyroscope. The upper magnet is molded so that the difference between its south pole and each of the lower magnets is significantly greater than the distance between its north pole and each of the lower magnets. A disk made of non-ferromagnetic material is attached to the upper magnet to increase the gyroscopic effect. Figure 12-13 illustrates this scheme. (The size of the upper magnet is exaggerated here for clarity.) The upper magnet is a spinning top that "balances" on the magnetic fields produced by itself and the lower array of magnets. A toy called the *Levitron*® provides a fascinating display of this principle. A good explanation of how it works can be found on the Internet at *www.levitron.com*.

A system, such as the one shown in Fig. 12-13, does not violate Earnshaw's Theorem. Remember that this theorem applies only to fixed magnets of constant intensity, such as those shown in the arrangements in Fig. 12-10. Nevertheless, the system shown in Fig. 12-13 is temperamental. The spin rate must be between certain limits, and the spinning magnet must be shaped just right. After awhile, air resistance will cause the spinning magnet to slow down to the point where the gyroscopic effect fails, and it flips over and sticks to one of the magnets in the lower array.

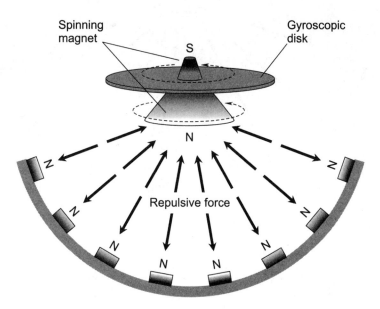

Fig. 12-13.   A spinning magnet equipped with a gyroscopic disk can levitate above an array of fixed magnets. (The size of the upper magnet is exaggerated for clarity.)

OSCILLATING FIELDS

An object that conducts electricity, but that is non-ferromagnetic, behaves as a diamagnetic material in the presence of an alternating (or oscillating) magnetic field. An oscillating magnetic field can be produced by a set of electromagnets to which high-frequency AC, rather than DC, is applied. A suitably shaped, rotating disk made of a material such as aluminum will levitate above an array of such electromagnets, as shown in Fig. 12-14.

This method of levitation works because of *eddy currents* that appear in the conducting disk. The eddy currents produce a secondary magnetic field that opposes the primary, oscillating field set up by the array of fixed AC electromagnets. This tends to expel the external magnetic flux from the disk. The disk becomes an "antimagnet." The rotation provides a stabilizing, gyroscopic effect, so the disk doesn't flip over. The disk stays centered as long as it is placed at along the central axis of the array to begin with, and as long as there isn't a significant disturbance to throw it off center.

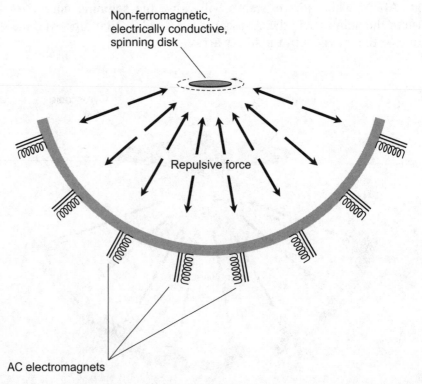

Fig. 12-14.   A spinning disk, made of non-ferromagnetic material that conducts electric currents, can levitate above a set of AC electromagnets.

Eventually the rotation rate of the disk will decay, because of air resistance, to the point that the system becomes unstable. This problem can be eliminated by placing the entire system in a vacuum. (The same thing can be done with the rotating-magnet system described in the previous section.) This allows the system to operate forever in theory, although in practice it will eventually fail because of inevitable system energy losses.

THE MAGLEV TRAIN

One of the most talked-about applications of magnetic levitation is the high-speed rail transit. Some passenger trains for urban commuters employ this technology, which is called *maglev*. The diamagnetic effects of super-conductors are most often used for such systems. An adaptation of the rotation scheme, described above, has also been tested on a small scale.

The chief advantage of maglev trains over the conventional trains is the fact that in a maglev system, the only friction is between the moving carriages and the air. There is no contact between the train and the track. The train hovers over the track. There is a gap of 2 to 3 centimeters (about an inch) between the train and the track. The track is a monorail. The train cars can be supported by either of two geometries, shown in Fig. 12-15. In either arrangement, vertical magnetic fields cause the train cars to be suspended above the track, and horizontal magnetic fields stabilize the cars so they remain centered. The forward or reverse motion is provided by devices called *linear motors*, which require an additional set of electromagnets in the track.

Superconductor type maglev systems rely on numerous, powerful electro-magnets embedded in the track to obtain the levitation. This is expensive, and raises the problem of shielding passengers from the strong magnetic fields. Magnetic shielding adds weight and expense to the construction of the train cars. Another potential problem is the ever-present danger of power outages. What will happen if a superconductor type maglev train traveling at 320 kilometers per hour (200 miles per hour), a speed typical of such systems, suddenly loses its support system and the cars settle onto the track? Yet another concern is foul weather. What will happen, for example, in a sudden snowstorm, windstorm, or hailstorm?

An alternative system called *Inductrack* uses permanent magnets in the cars and wire loops in the track. The motion of the cars with respect to the track produces the levitation, in a manner similar to the way a rotating, conducting disk levitates above a set of fixed magnets. This system travels on sets of small wheels as it first gets going. Once it is in motion at a few

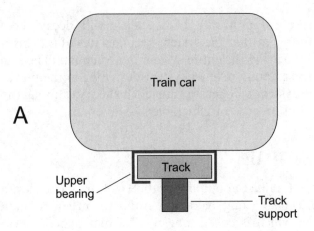

A

Train motion is perpendicular to page

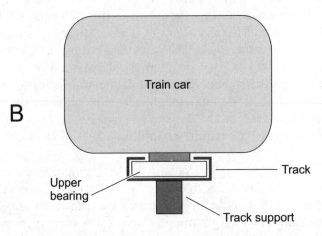

B

Fig. 12-15.   Simplified cross-sectional diagrams of maglev train geometries. At A, an upper bearing, attached to the car, wraps around and levitates above a monorail track. At B, the upper bearing levitates inside a wrap-around track. In both illustrations, train motion is perpendicular to the page.

kilometers per hour, the currents in the loops become sufficient to set up magnetic fields that repel the permanent magnets in the train cars. As with the superconductor type maglev train, the Inductrack system uses linear motors to achieve propulsion. If the power fails, the cars coast to a stop, but they do not fall onto the track suddenly at high speed. They settle onto the wheels, once the speed drops below a few kilometers per hour.

Maglev trains are controversial. They are capable of higher speeds than the conventional trains, but are expensive, and the technology is complex. Some engineers say that the benefits of maglev trains do not outweigh the high cost and technical difficulties. Proponents cite the potential for lower noise, reduced commute times, and the use of non-polluting energy sources.

[?] **PROBLEM 12-10**

How can passengers be shielded from the strong magnetic fields in a super-conductor type maglev train?

[✔] **SOLUTION 12-10**

The train cars can be made of a ferromagnetic substance such as steel, and care can be taken to ensure that the enclosure surrounds the passenger compartments sufficiently to keep the magnetic flux out. Unfortunately, steel is much heavier than aluminum, the other metal commonly used in general construction. Aluminum is not ferromagnetic and would offer no protection against magnetic fields unless it were made to carry high and potentially dangerous electric currents.

[?] **PROBLEM 12-11**

How can a maglev train negotiate a steep grade? Won't it fall downhill and settle at the bottom of a valley if there is no friction to provide braking action?

[✔] **SOLUTION 12-11**

The linear motors used in maglev train systems can drive the cars up steeper grades than is possible with the conventional trains. In addition, the linear motors can provide the braking action by switching into reverse, and they can keep the train from falling downgrade by operating against the force of gravity.

Quiz

This is an "open book" quiz. You may refer to the text in this chapter. A good score is 8 correct answers. Answers are in Appendix 1.

1. A material that completely expels magnetic flux is said to be

   (a) perfectly ferromagnetic.
   (b) perfectly supermagnetic.
   (c) perfectly nonmagnetic.
   (d) perfectly diamagnetic.

2. A substance with high retentivity is good for making

   (a) a superconductor.
   (b) an AC electromagnet.
   (c) a permanent magnet.
   (d) a diamagnet.

3. A device that periodically reverses the magnetic field polarity to keep a DC motor rotating is

   (a) a commutator.
   (b) an armature coil.
   (c) a linear motor.
   (d) a field coil.

4. An advantage of a magnetic disk, as compared with magnetic tape, for data storage and retrieval is the fact that

   (a) a disk lasts longer.
   (b) data can be stored to, and retrieved from, a disk more quickly.
   (c) disks look better.
   (d) disks are less susceptible to external magnetic fields.

5. A speaker that converts AC into sound waves using a coil and magnet is an example of

   (a) a transducer.
   (b) a synchro.
   (c) a superconductor.
   (d) a commutator.

6. Refer to Fig. 12-9. Suppose you are playing a tape back (that is, listening to data that is contained on an audio tape). Suddenly, the audio amplifier between the microphone and the recording head stops working. Nothing else in the system changes. What will you notice immediately?

    (a) Nothing. The tape will keep playing back normally.
    (b) The volume of the played-back sound will decrease.
    (c) Sound will stop coming from the speaker.
    (d) You'll hear the sounds from two different tape tracks coming out of the speaker at the same time.

7. Eddy currents in an electrical conductor produce

    (a) magnetic fields.
    (b) ferromagnetism.
    (c) retentivity.
    (d) All of the above

8. Suppose you want to make a system for the wind vane on your roof, allowing you to read the wind direction from a circular, compass-like scale in your living room. What type of device might you use?

    (a) A relay.
    (b) A selsyn.
    (c) A solenoid.
    (d) A fluxgate magnetometer.

9. Earnshaw's Theorem applies only to systems of

    (a) rotating magnets.
    (b) AC electromagnets.
    (c) stationary magnets.
    (d) levitating magnets.

10. Suppose you wind a length of wire into a coil on a cardboard cylinder left over from a used-up roll of paper towels. You place a rod, made of material having $\mu = 1$, inside the cylinder. What will happen to the magnetic flux inside the cylinder?

    (a) It will be expelled.
    (b) Its density will increase.
    (c) Its density will decrease.
    (d) Its density will not change.

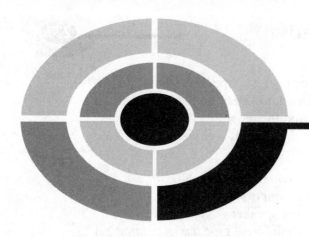

Test: Part Three

Do not refer to the text when taking this test. You may draw diagrams or use a calculator, if necessary. A good score is at least 30 answers (75% or more) correct. Answers are in Appendix 1. It's best to have a friend check your score the first time, so you won't memorize the answers, if you want to take the test again.

1. In order to generate an EM field, charge carriers must be

    (a) stationary.
    (b) moving.
    (c) accelerating.
    (d) electrically negative.
    (e) electrically positive.

2. The magnetic permeability of common substances is never

    (a) more than 1.
    (b) less than 1.
    (c) more than is the case in free space.
    (d) less than is the case in free space.
    (e) negative.

3. Suppose you are at the geographic north pole, that is, at latitude 90°. From this point of view, when you sight along the horizon, geomagnetic north corresponds to

    (a) geographic north.
    (b) geographic south.
    (c) geographic east.
    (d) geographic west.
    (e) any and all directions.

4. One of the major differences between a liquid crystal display (LCD) and a cathode-ray-tube (CRT) display is the fact that

    (a) the LCD costs less for the same screen size.
    (b) the LCD weighs more for the same screen size.
    (c) the CRT produces more ELF energy than the LCD.
    (d) the LCD can be used with computers, but the CRT cannot.
    (e) the CRT has a flat panel, but the LCD does not.

5. Electronic semiconductor switches are preferable to relays in most situations because

    (a) they can perform more switching operations per second than relays.
    (b) they can withstand higher currents than relays.
    (c) they can withstand higher voltages than relays.
    (d) they are more physically rugged than relays.
    (e) —Forget it! Relays are superior to electronic switches in every way.

6. In Fig. Test3-1, the object marked X is an example of

    (a) a monopole.
    (b) a dipole.
    (c) a tripole.
    (d) a virtual pole.
    (e) a geomagnetic pole.

7. In the situation shown by Fig. Test3-1, the flux density

    (a) is greatest near the object marked X.
    (b) is greatest far away from the object marked X.
    (c) does not vary with distance from the object marked X.
    (d) is a set of concentric circles, perpendicular to the dashed lines at every point.

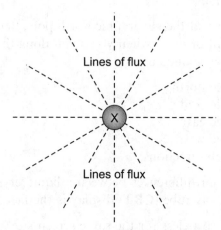

Fig. Test3-1.   Illustration for Part Three Test Questions 6 and 7.

   (e)   is a set of straight lines, passing through the plane of the page at
         right angles.

8.   A galvanometer can detect electric current because of

   (a)   the electromagnetic effect.
   (b)   the high resistance of the coil.
   (c)   changes in the geomagnetic field.
   (d)   the voltage difference around the coil.
   (e)   electromotive force inside the coil.

9.   Fill in the blank in the following sentence to make it true: "A
     _____ is an indicating device that uses stepper motors to show
     the direction in which an object is pointing."

   (a)   field coil
   (b)   compass
   (c)   solenoid
   (d)   relay
   (e)   selsyn

10.   Suppose two rod-shaped permanent magnets, with poles at the ends of
      the rods, are placed in close proximity, with their south poles facing
      each other. The rods are then brought together until their south-
      pole ends are in direct contact. What is observed between the two
      magnets as they get closer to each other?

   (a)   A decreasing repulsive force.
   (b)   A decreasing attractive force.

(c) An increasing repulsive force.
(d) An increasing attractive force.
(e) No force.

11. Earnshaw's theorem states that it is impossible to obtain levitation with sets of

   (a) permanent, stationary magnets.
   (b) moving electromagnets.
   (c) superconductors.
   (d) rotating disks and magnets.
   (e) magnets of any kind, in any system.

12. Solar flares are known to cause

   (a) geomagnetic storms.
   (b) hurricanes and tornadoes.
   (c) magnetic declination.
   (d) magnetic inclination.
   (e) All of the above

13. When electromagnetic interference to home entertainment equipment originates from a wireless transmitter such as a broadcast station, it is called

   (a) radio-frequency interference.
   (b) cellular interference.
   (c) hi-fi interference.
   (d) television interference.
   (e) wireless interference.

14. The formula for frequency f, in hertz, as a function of the wavelength λ, in meters, for an EM field in free space is:

$$f = 3 \times 10^8 / \lambda$$

What is the frequency of an EM field with a wavelength of 100 m?

   (a) 30 kHz
   (b) 300 kHz
   (c) 3 MHz
   (d) 30 MHz
   (e) 300 MHz

15. An electric current in a loop of superconducting wire can circulate

   (a) only as long as voltage is applied.

(b) only as long as the resistance of the loop remains infinite.
(c) only if the electron orbits in the atoms are all aligned.
(d) only if the temperature is sufficiently high.
(e) forever, in theory.

16. Which of the following statements (a), (b), (c), or (d), if any, is known to be true concerning extremely low frequency (ELF) fields?

    (a) When an object is exposed to an ELF field for a long enough time, that object will become radioactive.
    (b) Some of the atoms in human tissue exposed to ELF fields can become ionized, just as is the case with X rays and gamma rays.
    (c) Exposure to ELF fields can cause electrocution.
    (d) Exposure to the ELF from a computer causes blindness in laboratory rats.
    (e) None of the above statements (a), (b), (c), or (d) are true.

17. Magnetic lines of flux

    (a) are perpendicular to the direction of the magnetic field at every point in space.
    (b) are perpendicular to the axis of a magnet at every point in space.
    (c) are concentric circles that converge midway between two magnetic poles.
    (d) are parallel to the direction of the magnetic field at every point in space.
    (e) are parallel to the axis of a magnet at every point in space.

18. A disadvantage of maglev trains is the fact that

    (a) they require alternative energy sources to provide the electricity.
    (b) they cannot travel as fast as conventional trains.
    (c) they cost more to build and maintain than conventional trains.
    (d) they are inherently unstable because of the Meissner Effect.
    (e) Forget it! Maglev trains are superior in every way to conventional trains.

19. There is a limit to the flux density that can be generated in the core of an electromagnet. As the current in the coil increases, the flux density

inside the core goes up in proportion for a while. Beyond a certain point, the flux density levels off, and further increases in current do not produce any further increase in the flux density. This condition is known as

(a) permeability limitation.
(b) cutoff.
(c) remanence.
(d) core saturation.
(e) Meissner effect.

20. The inclination of the geomagnetic field is greatest

(a) at the geographic equator.
(b) at the geomagnetic equator.
(c) at the geographic poles.
(d) at the geomagnetic poles.
(e) along a band around the earth passing through both geomagnetic poles.

21. Magnetic levitation can be achieved with fixed magnets by means of

(a) the Earnshaw Effect.
(b) the use of materials with high resistance.
(c) the use of static, permanent magnets.
(d) a feedback system.
(e) a Meissner system.

22. The wavelength λ, in meters, as a function of the frequency f, in hertz, for an EM field in free space is given by the following formula:

$$\lambda = 3 \times 10^8 / f$$

What is the wavelength of an EM field that has a frequency of 600 kHz?

(a) 5 m
(b) 18 m
(c) 50 m
(d) 180 m
(e) None of the above

23. Suppose a galvanometer coil carries 20 mA of DC, causing the needle to be deflected 30° to the west of geomagnetic north. What will the needle do if the direction of the current remains the same, but the

intensity of the current is reduced to 10 mA? Assume that nothing else changes.

(a) It will be deflected to the west of geomagnetic north by more than 30°.
(b) It will be deflected to the west of geomagnetic north by less than 30°.
(c) It will be deflected to the east of geomagnetic north by more than 30°.
(d) It will be deflected to the east of geomagnetic north by less than 30°.
(e) It will point directly toward geomagnetic north.

24. If you hold your right hand with the thumb pointing out straight and the fingers curled, and then point your thumb in the direction of the conventional current flow in a straight wire, your fingers curl in the direction of the

(a) theoretical current.
(b) potential difference.
(c) geomagnetic field.
(d) magnetic lines of flux.
(e) magnetic inclination.

25. Figure Test3-2 is a logarithmic scale showing free-space EM field wavelengths, with the longest waves at the top and the shortest waves at the bottom. On the left-hand side of the scale, some common types of energy are listed. On the right-hand side, wavelengths in meters are shown, along with five points labeled A through E. Which point corresponds to a frequency of approximately 3 MHz?

(a) Point A
(b) Point B
(c) Point C
(d) Point D
(e) Point E

26. In Fig. Test3-2, point A corresponds to a frequency of approximately

(a) 300 Hz.
(b) 30 kHz.
(c) 3 MHz.
(d) 300 MHz.
(e) 30 GHz.

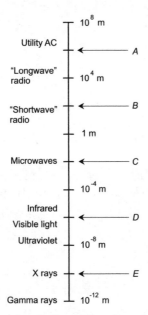

Fig. Test3-2.   Illustration for Part Three Test Questions 25 and 26.

27. The measure of how well a substance can "memorize" a magnetic field imposed on it, and thereby become a permanent magnet, is called

(a) permeability.
(b) reactance.
(c) retentivity.
(d) galvanism.
(e) Earnshaw effect.

28. The earth's magnetic poles are also called

(a) the electromagnetic poles.
(b) the geomagnetic poles.
(c) the virtual poles.
(d) the convergent poles.
(e) the astronomical poles.

29. Suppose a rod-shaped metal core is surrounded by a coil to make an electromagnet. The magnetic flux density in the core can be as great as 1000 G. Further increases in current cause no further increase in the flux density in the core. When the current is shut off, the flux

density in the core drops to 50 G. What is the retentivity of this core material?

(a)   1%
(b)   5%
(c)   20
(d)   50,000
(e)   It is impossible to determine this without more information.

30.   Fill in the following sentence to make it true: "In free space, the _____ of an EM field is inversely proportional to the wavelength."

(a)   intensity
(b)   period
(c)   frequency
(d)   flux density
(e)   orientation

31.   Which of the following has the shortest wavelength in free space?

(a)   ELF radiation.
(b)   Utility AC.
(c)   Visible light.
(d)   Microwave radio signals.
(e)   Shortwave radio signals.

32.   Figure Test3-3 is a simplified functional diagram of a DC motor. Component X is

(a)   the commutator.
(b)   the relay coil.
(c)   the armature coil.
(d)   the field coil.
(e)   the transformer.

33.   In Fig. Test3-3, the components labeled Y constitute

(a)   the commutator.
(b)   the relay coil.
(c)   the armature coil.
(d)   the field coil.
(e)   the transformer.

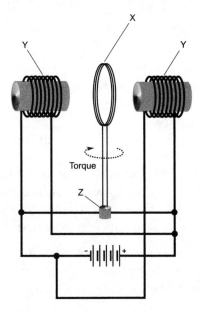

Fig. Test3-3. Illustration for Part Three Test Questions 32 through 34.

34. In Fig. Test3-3, the component labeled Z is

   (a) the commutator.
   (b) the relay coil.
   (c) the armature coil.
   (d) the field coil.
   (e) the transformer.

35. A ferromagnetic material

   (a) is repelled from a permanent magnet.
   (b) is not affected by a permanent magnet.
   (c) is attracted to a permanent magnet.
   (d) dilates magnetic lines of flux.
   (e) becomes electrically charged in the vicinity of a magnet.

36. Solar wind is another name for

   (a) geomagnetic storms that occur in the earth's atmosphere.
   (b) hurricanes and tornadoes that are caused by solar eruptions.
   (c) magnetic declination.
   (d) magnetic inclination.
   (e) the stream of subatomic particles emanating from the sun.

37. In a home entertainment system, the installation of an electrical ground

    (a) increases the susceptibility of the equipment to problems associated with electromagnetic interference (EMI).
    (b) increases the risk of electric shock or electrocution to people using the equipment.
    (c) increases the risk of overloading the microprocessor chips in individual devices.
    (d) cannot be accomplished without an extensive system of ground wires and radials, along with a special fuse box.
    (e) can help to minimize the susceptibility of the equipment to problems associated with electromagnetic interference (EMI).

38. Suppose a galvanometer coil carries 20 mA of DC, causing the needle to be deflected 30° to the west of geomagnetic north. What will the needle do if the current is cut off? Assume nothing else changes.

    (a) It will be deflected to the west of geomagnetic north by more than 30°.
    (b) It will be deflected to the west of geomagnetic north by less than 30°.
    (c) It will be deflected to the east of geomagnetic north by more than 30°.
    (d) It will be deflected to the east of geomagnetic north by less than 30°.
    (e) It will point directly toward geomagnetic north.

39. Which of the following has the lowest frequency?

    (a) Ultraviolet rays.
    (b) Infrared rays.
    (c) Gamma rays.
    (d) Visible light.
    (e) Radio waves.

40. Fill in the blank to make the following sentence true: "The greatest magnetic field strength around a rod-shaped permanent magnet is near the _____, where the flux lines converge or diverge."

    (a) center
    (b) poles
    (c) axis
    (d) cross section
    (e) surface

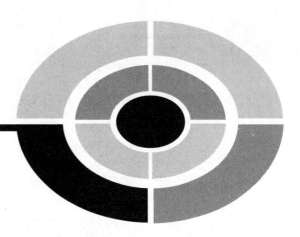

Final Exam

Do not refer to the text when taking this test. You may draw diagrams or use a calculator, if necessary. A good score is at least 53 answers (75% or more) correct. Answers are in Appendix 1. It's best to have a friend check your score the first time, so that you won't memorize the answers, if you want to take the test again.

1. The current, voltage, or power level of an AC or a fluctuating wave at a specific point in time is called the

    (a) rms amplitude.
    (b) instantaneous amplitude.
    (c) time-point amplitude.
    (d) zero-point amplitude.
    (e) effective amplitude.

2. Which, if any, of the following particles (a), (b), (c), or (d) is not an electric charge carrier?

    (a) An atomic nucleus
    (b) An electron
    (c) A proton
    (d) A neutron
    (e) All of the above (a), (b), (c), and (d) carry electric charge.

3. A common method of defining the phase of an AC cycle is to divide it into 360 equal parts called

    (a)   root mean squares.
    (b)   radians.
    (c)   phases.
    (d)   degrees.
    (e)   increments.

4. How many cells, each supplying 1.5 V, must be connected in series to obtain a battery that supplies 12 V?

    (a)   It is impossible to get a 12 V battery by connecting 1.5 V cells in series.
    (b)   2
    (c)   4
    (d)   8
    (e)   16

5. Suppose two AC waves are exactly in phase. Call them wave X and wave Y. There is no DC component on either wave. The peak-to-peak voltage of wave X is 60 V, and the peak-to-peak voltage of wave Y is 80 V. What is the positive peak voltage of the composite wave, $Z = X + Y$?

    (a)   +60 V
    (b)   +70 V
    (c)   +80 V
    (d)   +140 V
    (e)   More information is needed in order to determine this.

6. If thunder can be heard, you can be certain that

    (a)   lightning is occurring within a few kilometers of your location, whether or not it is visible.
    (b)   lightning is occurring within a few kilometers of your location, but only if it is visible.
    (c)   lightning is occurring, but it is a long way off (more than a few kilometers away from your location).
    (d)   lightning is about to strike you.
    (e)   lightning is taking place, but only from cloud to cloud.

7. Fill in the blank to make the following sentence true: "In a typical electrical conductor such as copper or aluminum wire, the _____ can move easily from atom to atom."

   (a) nuclei
   (b) electrons
   (c) photons
   (d) protons
   (e) neutrons

8. Figure Exam-1 shows a set of 4 identical components connected in

   (a) parallel.
   (b) series.
   (c) series-parallel.
   (d) cascade.
   (e) inverse.

9. Suppose each of the components in Fig. Exam-1 supplies 0.6 V. What is the voltage E?

   (a) 0 V
   (b) 0.15 V
   (c) 0.6 V
   (d) 2.4 V
   (e) It is impossible to answer this without more information.

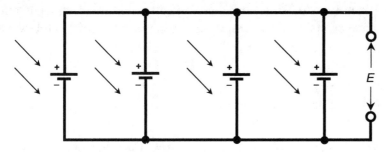

Fig. Exam-1. Illustration for Final Exam Questions 8 through 12.

10. Suppose each of the components in Fig. Exam-1 delivers 50 mA. What is the total current delivered by the combination at the output terminals?

    (a) 0 mA
    (b) 12.5 mA
    (c) 50 mA
    (d) 200 mA
    (e) It is impossible to answer this without more information.

11. Suppose each of the components in Fig. Exam-1 delivers 30 mW. What is the total power delivered by the combination at the output terminals?

    (a) 0 mW
    (b) 7.5 mW
    (c) 30 mW
    (d) 120 mW
    (e) It is impossible to answer this without more information.

12. Each of the 4 interconnected components shown in Fig. Exam-1 is

    (a) an alkaline cell.
    (b) a photovoltaic cell.
    (c) a fuel cell.
    (d) an electromagnet.
    (e) an AC generator.

13. When two AC sine waves have exactly the same frequency and their wave crests occur at exactly the same points in time, the waves are

    (a) out of phase.
    (b) in phase opposition.
    (c) in phase coincidence.
    (d) in phase quadrature.
    (e) in phase inversion.

14. What is an advantage of half-wave rectification over full-wave rectification in a power supply?

    (a) The output waveform of a half-wave rectifier is easier to filter.
    (b) The output voltage from a half-wave rectifier is more constant if the current demand varies greatly.
    (c) Half-wave rectification puts less strain on the transformer.
    (d) Half-wave rectification puts less strain on the diodes.
    (e) None of the above

15. Power is a measure of

    (a) voltage increase per unit time.
    (b) charge carrier flow per unit time.
    (c) resistance change per unit time.
    (d) energy consumption per unit time.
    (e) current consumption per unit time.

16. Wire can carry a little more than the maximum rated continuous current for brief moments. But sustained, excessive current

    (a) can cause the wire to soften.
    (b) produces overheating.
    (c) is dangerous because electrical fires can result.
    (d) makes the wire abnormally susceptible to breakage.
    (e) All of the above

17. Which of the following statements is false, with regard to DC circuits?

    (a) The resistance of a component is equal to the negative of the conductance of that component.
    (b) The current through a component having a fixed resistance is directly proportional to the voltage across it.
    (c) The voltage across a component having a fixed resistance is directly proportional to the current through it.
    (d) The power dissipated by a component having a fixed resistance is directly proportional to the square of the voltage across it.
    (e) The power dissipated by a component having a fixed resistance is directly proportional to the square of the current through it.

18. The British Standard Wire Gauge (NBS SWG), American Wire Gauge (AWG), and Birmingham Wire Gauge (BWG) all share a common characteristic. What is it?

    (a) They all specify wire size as a number that increases as the cross-sectional area increases.
    (b) They all specify wire size as a number that increases as the diameter increases.
    (c) They all specify wire size in terms of the conductance in siemens.
    (d) They all specify wire size as a number that increases as the resistivity increases.
    (e) They all specify the conductivity of a span of wire as a function of its length.

19. The conductivity of uninsulated, solid copper wire is directly proportional to

   (a) the size in AWG.
   (b) the size in NBS SWG.
   (c) the size in BWG.
   (d) the diameter.
   (e) the cross-sectional area.

20. Suppose 6 C of electric charge carriers flow past a given point in 2 s. What is the current?

   (a) 12 A
   (b) 8 A
   (c) 6 A
   (d) 3 A
   (e) 2 A

21. A magnetic field is always produced by

   (a) stationary electrons.
   (b) moving electrons.
   (c) stationary atoms.
   (d) a potential difference.
   (e) Any of the above

22. Suppose a certain point in a circuit has 10 branches leading into it, each carrying 100 mA of current. There are 2 branches leading out of this point. What can we say about the currents in the outgoing branches?

   (a) Both of the outgoing branches carry 1 A of current.
   (b) Both of the outgoing branches carry 500 mA of current.
   (c) The sum of the currents in the outgoing branches is 1 A.
   (d) The sum of the currents in the outgoing branches is 500 mA.
   (e) Nothing at all, without further information.

23. Figure Exam-2 is a graph or display of

   (a) a sine wave.
   (b) a ramp wave.
   (c) a sawtooth wave.
   (d) an elliptical wave.
   (e) an asymmetrical wave.

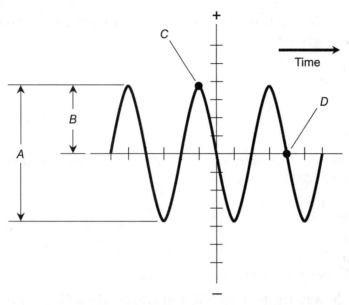

Fig. Exam-2.   Illustration for Final Exam Questions 23 through 27.

24.   Which of the labeled points or spans in Fig. Exam-2 shows a wave trough?

(a)   *A*
(b)   *B*
(c)   *C*
(d)   *D*
(e)   None of them

25.   Which of the labeled points or spans in Fig. Exam-2 shows the peak-to-peak amplitude?

(a)   *A*
(b)   *B*
(c)   *C*
(d)   *D*
(e)   None of them

26. Which of the labeled points or spans in Fig. Exam-2 shows a positive-going zero point?

(a) A
(b) B
(c) C
(d) D
(e) None of them

27. Which of the labeled points or spans in Fig. Exam-2 shows the rms amplitude?

(a) A
(b) B
(c) C
(d) D
(e) None of them

28. If the wavelength of an EM field increases by a factor of 100, the frequency

(a) becomes 1/100 as great.
(b) becomes 1/10 as great.
(c) does not change.
(d) becomes 10 times as great.
(e) becomes 100 times as great.

29. Figure Exam-3 is a schematic diagram of

(a) a power supply without a rectifier.
(b) a power supply with a full-wave, center-tap rectifier.
(c) a power supply with a full-wave bridge rectifier.
(d) a voltage-doubler power supply.
(e) a voltage regulator.

30. In the circuit of Fig. Exam-3, what is the function of the component labeled P?

(a) It acts as a rectifier.
(b) It converts DC to AC.
(c) It steps up the AC voltage.
(d) It helps to filter the rectifier output.
(e) None of the above

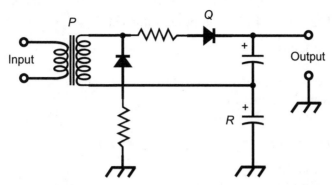

Fig. Exam-3. Illustration for Final Exam Questions 29 through 32.

31. In the circuit of Fig. Exam-3, what is the function of the component labeled Q?

    (a) It acts as a rectifier.
    (b) It converts DC to AC.
    (c) It steps up the AC voltage.
    (d) It helps to filter the rectifier output.
    (e) None of the above

32. In the circuit of Fig. Exam-3, what is the function of the component labeled R?

    (a) It acts as a rectifier.
    (b) It converts DC to AC.
    (c) It steps up the AC voltage.
    (d) It helps to filter the rectifier output.
    (e) None of the above

33. Suppose you have an unlimited supply of resistors, each having a resistance of $50\,\Omega$ and capable of dissipating up to $1\,\mathrm{W}$. You need a resistor that has a value of $50\,\Omega$, but it must be capable of dissipating $7\,\mathrm{W}$ or more. What can you do to obtain such a resistor?

    (a) Connect 7 of the $50\,\Omega$ resistors in series.
    (b) Connect 7 of the $50\,\Omega$ resistors in parallel.
    (c) Connect 4 of the $50\,\Omega$ resistors in a 2-by-2 series-parallel matrix.
    (d) Connect 9 of the $50\,\Omega$ resistors in a 3-by-3 series-parallel matrix.
    (e) Nothing. It's impossible to do this.

34. In Fig. Exam-4, the component labeled *V* is

    (a) a lamp.
    (b) a resistor.
    (c) a potentiometer.
    (d) an AC generator.
    (e) None of the above

35. In Fig. Exam-4, the component labeled *W* is

    (a) a lamp.
    (b) a motor.
    (c) a switch.
    (d) a potentiometer.
    (e) None of the above

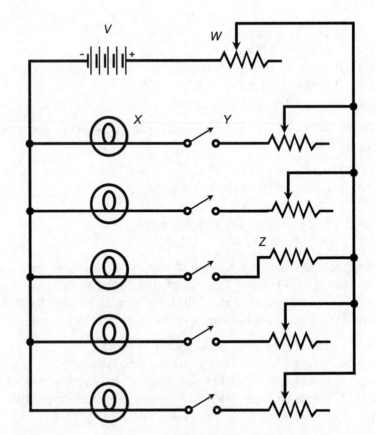

Fig. Exam-4. Illustration for Final Exam Questions 34 through 39.

36.   In Fig. Exam-4, the component labeled X is

(a)   a lamp.
(b)   a switch.
(c)   an AC generator.
(d)   a solar cell.
(e)   None of the above

37.   In Fig. Exam-4, the component labeled Y is

(a)   a motor.
(b)   a relay.
(c)   a switch.
(d)   a battery.
(e)   None of the above

38.   In Fig. Exam-4, the component labeled Z is

(a)   a resistor.
(b)   a battery.
(c)   a cell.
(d)   a lamp.
(e)   None of the above

39.   In Fig. Exam-4, the condition, setting, or value of component Y affects the condition of

(a)   component V.
(b)   component W.
(c)   component X.
(d)   all of the components V, W, and X.
(e)   none of the components V, W, or X.

40.   An ideal battery delivers a constant current for awhile, and then the current starts to decrease. When the current output from such a battery is plotted as a function of time, the resulting graph is called

(a)   a flat discharge curve.
(b)   an increasing discharge curve.
(c)   a gradual discharge curve.
(d)   a declining discharge curve.
(e)   a precipitous discharge curve.

41. Suppose an electric bulb designed for use at 12 V DC has constant resistance, no matter how much current it carries (as long as the current is not so high that it burns the bulb out). Suppose that the current through the bulb is 4 A at 12 V DC. How much current passes through the bulb if it is supplied with 3 V DC?

    (a) 16 A
    (b) 1.33 A
    (c) 1 A
    (d) 0.75 A
    (e) More information is necessary to answer this.

42. Suppose the fuse in a power supply blows out. It is a 10 A, quick-break fuse. You replace it with a 10 A, slow-blow fuse. What, if anything, is wrong with doing this?

    (a) If there is a short circuit in the equipment connected to the supply, there is a risk that the slow-blow fuse will allow excessive current to flow for too long before it blows out.
    (b) The slow-blow fuse will blow out too quickly, if there is a temporary current surge, so you might have to replace the fuse when it should not be necessary.
    (c) Whenever a slow-blow fuse is used to replace a quick-break fuse, the slow-blow fuse should be rated at only half the current (in this case 5 A).
    (d) Whenever a slow-blow fuse is used to replace a quick-break fuse, the slow-blow fuse should be rated at twice the current (in this case 20 A).
    (e) Nothing is wrong in doing this. In fact, it's a good idea to use a slow-blow fuse in place of a quick-break fuse having the same current rating.

43. The deflecting coils in an electromagnetic cathode-ray tube (CRT)

    (a) generate magnetic fields that exert force on the electrons traveling through the CRT, causing the electron beam to change direction.
    (b) generate electric fields that exert force on small magnets in the CRT, causing deflection of the electron beam.
    (c) produce electrons that are accelerated by the anodes, producing a visible spot on the CRT screen.
    (d) are positively charged, so they increase the speed of the electrons traveling through the CRT.

(e) are negatively charged, so they repel the electrons traveling through the CRT and thereby push them up, down, or sideways.

44. Suppose you coil a length of bell wire around a steel bolt, and then connect the coil to a source of DC that has variable voltage and can supply an unlimited amount of current. As you increase the voltage, the steel bolt becomes more and more magnetized until, at a certain point, further increases in the voltage do not cause any further increase in the magnetization of the bolt. At this point, the steel bolt is said to be in a state of

(a) permanent magnetization.
(b) core saturation.
(c) eddy-current loss.
(d) superconductivity.
(e) magnetic remanence.

45. Figure Exam-5 shows a current-carrying wire as viewed from right along its axis. The large, clear arrow marked K points directly at you, straight out of the page. The dashed circle represents

(a) the flow of current.
(b) the potential difference.
(c) an electric line of flux.
(d) a magnetic line of flux.
(e) None of the above

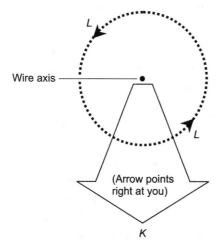

Fig. Exam-5. Illustration for Final Exam Questions 45 through 47.

46. In Fig. Exam-5, suppose the magnetic flux flows counterclockwise. The conventional current

   (a) follows the dashed circle in the direction of the arrows marked *L*.
   (b) follows the dashed circle, contrary to the direction of the arrows marked *L*.
   (c) points in the direction of the large, clear arrow marked *K*.
   (d) points contrary to the direction of the large, clear arrow marked *K*.
   (e) None of the above

47. In Fig. Exam-5, what do the arrows marked *L* represent?

   (a) The direction of the potential difference.
   (b) The direction of the electric flux.
   (c) The direction of the permeability.
   (d) The direction of the retentivity.
   (e) None of the above

48. The geomagnetic field

   (a) has an axis that does not correspond exactly with the earth's rotational axis.
   (b) has an axis that corresponds exactly with the earth's rotational axis.
   (c) is perfectly symmetrical all around the earth.
   (d) has lines of flux that flow only in the plane defined by the equator.
   (e) has poles that lie on the equator.

49. Imagine a "black box" with two terminals, called point *X* and point *Y*, on its exterior. The resistance between points *X* and *Y* can vary. A constant DC voltage is connected between points *X* and *Y*. Which of the following statements is true?

   (a) The number of amperes that flow through the box is inversely proportional to the number of ohms between points *X* and *Y*.
   (b) The number of amperes that flow through the box is directly proportional to the number of ohms between points *X* and *Y*.
   (c) The number of amperes that flow through the box does not change as the number of ohms between points *X* and *Y* varies.
   (d) The number of amperes that flow through the box is inversely proportional to the square of the number of ohms between points *X* and *Y*.

(e) The number of amperes that flow through the box is directly proportional to the square of the number of ohms between points X and Y.

50. A quantitative expression of the extent to which a substance concentrates magnetic lines of flux, relative to the flux density in air or a vacuum, is

(a) retentivity.
(b) conductivity.
(c) resistivity.
(d) remanence.
(e) None of the above

51. For an electromagnetic field to be generated, it is necessary for electric charge carriers to

(a) accumulate.
(b) dissipate.
(c) move fast.
(d) change velocity.
(e) travel in circles.

52. A fluxgate magnetometer can be used as part of

(a) a robot navigation system.
(b) an electric power plant.
(c) a radio transmitter.
(d) a computer display.
(e) a cathode-ray oscilloscope.

53. In theory, a magnetic field extends into space forever, unless something blocks it, such as

(a) a current-carrying wire.
(b) the solar wind.
(c) another magnetic field.
(d) an enclosure made of steel.
(e) an electric field.

54. The term *static electricity* refers to

    (a) an accumulation of charge carriers without a flow of current.
    (b) charge carriers moving in circles.
    (c) charge carriers moving at constant velocity.
    (d) accelerated charge carriers.
    (e) charge carriers that cause interference to radio reception.

55. As the power demanded from an AC electric generator increases, the mechanical force required to drive it

    (a) does not change.
    (b) increases.
    (c) decreases.
    (d) becomes negative.
    (e) becomes zero.

56. If a coil of insulated bell wire is wound around a magnetic compass, the resulting instrument can be used to

    (a) measure the intensity of the solar wind.
    (b) distort the geomagnetic field.
    (c) generate AC.
    (d) detect DC in the wire.
    (e) measure the intensity of an EM field.

57. Fill in the blank to make the following sentence true: "A _____ carries current or signals along several paths from one point to another, independently and at the same time."

    (a) parallel cable
    (b) stranded wire
    (c) lamp cord
    (d) serial cable
    (e) magnetic cable

58. The lines of flux in the vicinity of a magnetic dipole

    (a) are always straight lines.
    (b) are circles centered at the poles.
    (c) converge toward one pole and diverge from the other pole.
    (d) converge midway between the two poles.
    (e) Forget it! There is no such thing as a magnetic dipole.

59. A 2-wire utility electrical plug with unequal-width blades, commonly seen on cords for simple appliances such as desk lamps, is

   (a) a serial plug.
   (b) a parallel plug.
   (c) an electromagnetic plug.
   (d) a coaxial plug.
   (e) a polarized plug.

60. Suppose two AC sine waves are exactly in phase. Call them wave X and wave Y. There is no DC component on either wave. The peak-to-peak voltage of wave X is 50 V, and the peak-to-peak voltage of wave Y is 90 V. What is the peak-to-peak voltage of the composite wave, $Z = X + Y$?

   (a) 40 V
   (b) 50 V
   (c) 90 V
   (d) 140 V
   (e) More information is needed in order to determine this.

61. Figure Exam-6 is a simplified functional drawing of a dynamic transducer. The component labeled R is

   (a) a cylindrical piece of non-ferromagnetic material.
   (b) a permanent magnet.
   (c) a superconductor.
   (d) a levitating device.
   (e) a diaphragm.

62. In Fig. Exam-6, assuming the transducer is properly connected and operated, the component labeled S carries

   (a) DC at radio frequencies.
   (b) DC at audio frequencies.
   (c) AC at radio frequencies.
   (d) AC at audio frequencies.
   (e) magnetic flux.

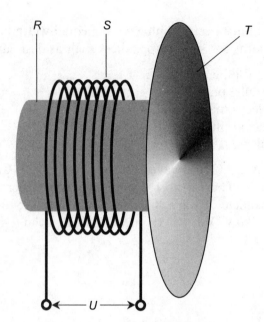

R   S   T

←— U —→

Fig. Exam-6. Illustration for Final Exam Questions 61 through 64.

63. What function does the component labeled *T* in Fig. Exam-6 perform?

(a) It converts AC to DC.
(b) It converts DC into mechanical vibrations.
(c) It converts DC to AC.
(d) It intercepts or generates sound waves.
(e) It intercepts or generates radio waves.

64. In Fig. Exam-6, assuming the transducer is properly connected and operated, what appears between the terminals marked *U*?

(a) A low AC voltage.
(b) A low DC voltage.
(c) A high DC voltage.
(d) Static electricity.
(e) A magnetic field.

65. The sum of all the voltages, as you go around any closed loop in a DC circuit from some fixed point and return there from the opposite direction, and taking polarity into account, is always equal to

    (a) the product of all the battery voltages.
    (b) the product of the resistance and the square of the current.
    (c) the resistance divided by the current.
    (d) the difference between the resistance and the current.
    (e) zero.

66. Refer to Fig. Exam-7. Suppose $E = 18$ V and $R = 180\,\Omega$. What is the meter reading?

    (a) 100 mA
    (b) 1.8 A
    (c) 10 A
    (d) 3240 A
    (e) It is impossible to answer this without more information.

67. Refer to Fig. Exam-7. Suppose $I = 200$ mA and $R = 120\,\Omega$. What is the battery voltage?

    (a) 36 kV
    (b) 600 V
    (c) 24 V
    (d) 1.67 V
    (e) It is impossible to answer this without more information.

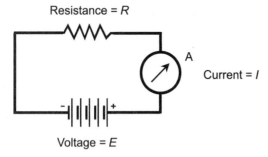

Fig. Exam-7. Illustration for Final Exam Questions 66 through 70.

68. Refer to Fig. Exam-7. Suppose $E = 6$ V and $I = 600\,\mu$A. What is the value of the resistor?

    (a) $10\,\text{k}\Omega$
    (b) $100\,\Omega$
    (c) $1\,\Omega$
    (d) $0.01\,\Omega$
    (e) It is impossible to answer this without more information.

69. Refer to Fig. Exam-7. Suppose $E = 12$ V and $I = 2$ A. What is the power dissipated by the resistor?

    (a) $167\,\text{mW}$
    (b) $333\,\text{mW}$
    (c) $6\,\text{W}$
    (d) $24\,\text{W}$
    (e) It is impossible to answer this without more information.

70. Refer to Fig. Exam-7. Suppose $E = 12$ V and $I = 2$ A. What is the energy demanded from the battery?

    (a) $2\,\text{Wh}$
    (b) $4\,\text{Wh}$
    (c) $76\,\text{Wh}$
    (d) $288\,\text{Wh}$
    (e) It is impossible to answer this without more information.

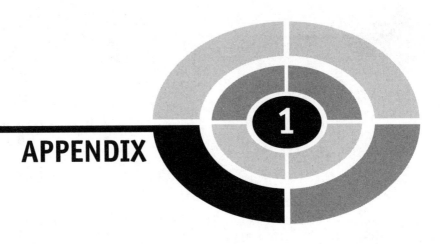

Answers to Quiz, Test, and Exam Questions

Chapter 1

1. b	2. a	3. c	4. a	5. d
6. b	7. a	8. a	9. d	10. c

Chapter 2

1. c	2. a	3. a	4. a	5. b
6. a	7. b	8. d	9. c	10. c

Chapter 3

1. b	2. b	3. a	4. b	5. a
6. c	7. d	8. d	9. c	10. d

Chapter 4

1. d	2. b	3. c	4. c	5. b
6. b	7. b	8. b	9. d	10. d

Chapter 5

1. d	2. a	3. c	4. a	5. d
6. a	7. c	8. c	9. b	10. b

Test: Part One

1. a	2. e	3. c	4. d	5. c
6. c	7. b	8. e	9. d	10. b
11. d	12. a	13. d	14. d	15. b
16. a	17. a	18. e	19. a	20. b
21. b	22. d	23. d	24. e	25. b
26. d	27. e	28. c	29. a	30. a
31. a	32. e	33. a	34. e	35. b
36. a	37. e	38. a	39. b	40. b

Chapter 6

1. b	2. a	3. a	4. b	5. b
6. d	7. c	8. d	9. a	10. d

Chapter 7

1. c	2. a	3. a	4. a	5. d

6. b   7. b   8. d   9. d   10. c

Chapter 8

1. c   2. d   3. a   4. c   5. b
6. b   7. c   8. a   9. d   10. a

Chapter 9

1. d   2. b   3. c   4. d   5. d
6. a   7. c   8. c   9. b   10. c

Test: Part Two

1. b	2. b	3. a	4. b	5. e
6. d	7. e	8. c	9. a	10. d
11. d	12. e	13. d	14. b	15. b
16. e	17. c	18. c	19. d	20. a
21. a	22. d	23. c	24. a	25. d
26. b	27. e	28. c	29. b	30. d
31. c	32. c	33. a	34. d	35. d
36. d	37. c	38. c	39. b	40. a

Chapter 10

1. c   2. a   3. b   4. a   5. d
6. c   7. d   8. c   9. d   10. b

Chapter 11

1. b   2. a   3. d   4. a   5. c
6. c   7. c   8. c   9. d   10. c

Chapter 12

1.d	2. c	3. a	4. b	5. a
6. a	7. a	8. b	9. c	10. d

Test: Part Three

1. c	2. e	3. b	4. c	5. a
6. a	7. a	8. a	9. e	10. c
11. a	12. a	13. a	14. c	15. e
16. e	17. d	18. c	19. d	20. d
21. d	22. e	23. b	24. d	25. b
26. a	27. c	28. b	29. b	30. c
31. c	32. c	33. d	34. a	35. c
36. e	37. e	38. e	39. e	40. b

Final Exam

1. b	2. d	3. d	4. d	5. b
6. a	7. b	8. a	9. c	10. d
11. d	12. b	13. c	14. e	15. d
16. e	17. a	18. d	19. e	20. d
21. b	22. c	23. a	24. e	25. a
26. e	27. e	28. a	29. d	30. c
31. a	32. d	33. d	34. e	35. d
36. a	37. c	38. a	39. c	40. a
41. c	42. a	43. a	44. b	45. d
46. c	47. e	48. a	49. a	50. e
51. d	52. a	53. d	54. a	55. b
56. d	57. a	58. c	59. e	60. d
61. b	62. d	63. d	64. a	65. e
66. a	67. c	68. a	69. d	70. e

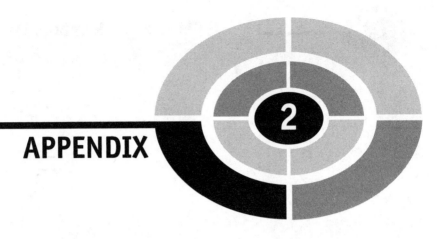
Symbols Used in Schematic Diagrams

ammeter

amplifier, general

amplifier, inverting

amplifier, operational

AND gate

antenna, balanced

antenna, general

antenna, loop

antenna, loop, multiturn

battery, electrochemical

capacitor, feedthrough

capacitor, fixed

capacitor, variable

capacitor, variable, split-rotor

capacitor, variable, spit-stator

cathode, electron-tube, cold

cathode, electron-tube,
directly heated

cathode, electron-tube,
indirectly heated

cavity resonator

cell, electrochemical

circuit breaker

coaxial cable

crystal, piezoelectric

delay line

diac

diode, field-effect

diode, general

diode, Gunn

diode, light-emitting

diode, photosensitive

diode, PIN

diode, Schottky

diode, tunnel

diode, varactor

diode, Zener

directional coupler

directional wattmeter

exclusive-OR gate

female contact, general

ferrite bead

filament, electron-tube

fuse

galvanometer	
grid, electron-tube	
ground, chassis	
ground, earth	
handset	
headset, double	
headset, single	
headset, stereo	
inductor, air core	
inductor, air core, bifilar	
inductor, air core, tapped	
inductor, air core, variable	
inductor, iron core	

inductor, iron core, bifilar

inductor, iron core, tapped

inductor, iron core, variable

inductor, powdered-iron core

inductor, powdered-iron core, bifilar

inductor, powdered-iron core, tapped

inductor, powdered-iron core, variable

or

integrated circuit, general

(Part No.)

jack, coaxial or phono

jack, phone, 2-conductor

jack, phone, 3-conductor

key, telegraph

lamp, incandescent

lamp, neon

male contact, general

meter, general

microammeter

microphone

microphone, directional

milliammeter

NAND gate

negative voltage connection

NOR gate

NOT gate

optoisolator

OR gate

outlet, 2-wire, nonpolarized

outlet, 2-wire, polarized

outlet, 3-wire

outlet, 234-volt

plate, electron-tube

plug, 2-wire, nonpolarized

plug, 2-wire, polarized

plug, 3-wire

plug, 234-volt

plug, coaxial or phono

plug, phone, 2-conductor

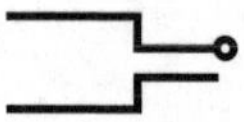

plug, phone, 3-conductor

positive voltage connection   +

potentiometer

probe, radio-frequency

or

rectifier, gas-filled

rectifier, high-vacuum

rectifier, semiconductor

rectifier, silicon-controlled

relay, double-pole, double-throw

relay, double-pole, single-throw

relay, single-pole, double-throw

relay, single-pole, single-throw

resistor, fixed

resistor, preset

resistor, tapped

resonator

rheostat

saturable reactor

signal generator

solar battery

solar cell

source, constant-current

source, constant-voltage

speaker

switch, double-pole, double-throw

switch, double-pole, rotary

switch, double-pole, single-throw

switch, momentary-contact

switch, silicon-controlled

switch, single-pole, double-throw

switch, single-pole, rotary

switch, single-pole, single-throw

terminals, general, balanced

terminals, general, unbalanced

test point

thermocouple

or

transformer, air core

transformer, air core, step-down

transformer, air core, step-up

transformer, air core, tapped primary

transformer, air core, tapped secondary

transformer, iron core

transformer, iron core, step-down

transformer, iron core, step-up

transformer, iron core, tapped primary

transformer, iron core, tapped secondary

transformer, powdered-iron core

transformer, powdered-iron core, step-down

transformer, powdered-iron core, step-up

transformer, powdered-iron core,
tapped primary

transformer, powdered-iron core,
tapped secondary

transistor, bipolar, NPN

transistor, bipolar, PNP

transistor, field-effect, N-channel

transistor, field-effect, P-channel

transistor, MOS field-effect, N-channel

transistor, MOS field-effect, P-channel

transistor, photosensitive, NPN

transistor, photosensitive, PNP

transistor, photosensitive, field-effect,
N-channel

transistor, photosensitive, field-effect,
P-channel

transistor, unijunction

triac

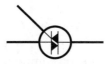

tube, diode

tube, heptode

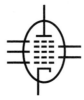

tube, hexode

tube, pentode

tube, photosensitive

tube, tetrode

tube, triode

unspecified unit or component

voltmeter

wattmeter

waveguide, circular

waveguide, flexible

waveguide, rectangular

waveguide, twisted

(preferred)

or

(alternative)

wires, crossing, connected

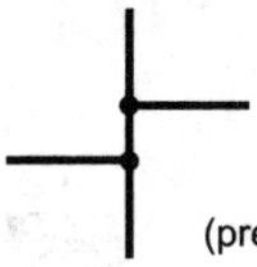

(preferred)

or

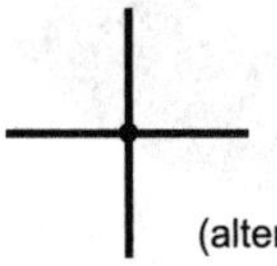

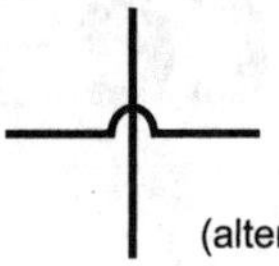

(alternative)

wires, crossing, not connected

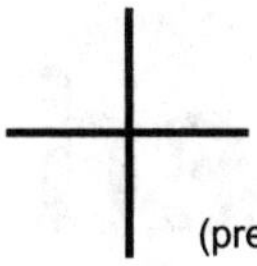

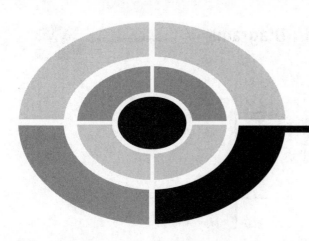

Suggested Additional References

Crowhurst, N. and Gibilisco, S., *Mastering Technical Mathematics*, 2nd edition (New York, NY: McGraw-Hill, 1999)

Dorf, R., *Electrical Engineering Handbook*, 2nd edition (Boca Raton, FL: CRC Press, 1997)

Gibilisco, S., *Electronics Demystified* (New York, NY: McGraw-Hill, 2005)

Gibilisco, S., *Teach Yourself Electricity and Electronics*, 3rd edition (New York, NY: McGraw-Hill, 2002)

Gottlieb, I. M., *Electric Motors and Control Techniques*, 2nd edition (New York, NY: TAB/McGraw-Hill, 1994)

Gussow, M., *Basic Electricity* (New York, NY: Schaum's Outlines, McGraw-Hill, 1983)

Morrison, R., *Electricity: A Self-Teaching Guide* (Hoboken, NJ: John Wiley & Sons, Inc., 2003)

Slone, G. R., *TAB Electronics Guide to Understanding Electricity and Electronics*, 2nd edition (New York, NY: McGraw-Hill, 2000)

Van Valkenburg, N., *Basic Electricity: Complete Course* (Albany, NY: Delmar Learning, 1995)

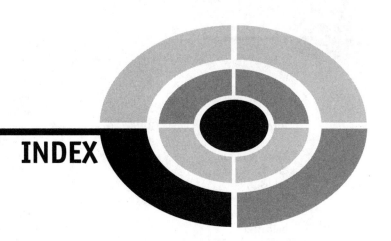

INDEX

ABOUT THE AUTHOR

Stan Gibilisco is one of McGraw-Hill's most prolific and popular authors. His clear, reader-friendly writing style makes his science, electronics, and mathematics books accessible to a wide audience. His background in mathematics, engineering, and research makes him an ideal editor for tutorials and professional handbooks. He is the author of *Teach Yourself Electricity and Electronics, The Illustrated Dictionary of Electronics*, and the *Electronics Portable Handbook*, among more than 20 other books and dozens of magazine articles. Booklist named his *McGraw-Hill Encyclopedia of Personal Computing* one of the "Best References of 1996."